# Crop Trials with R

NIPA® GENX ELECTRONIC RESOURCES & SOLUTIONS P. LTD.
New Delhi-110 034

## About the Authors

**Dr. Rumit Patel,** alumni of Anand Agricultural University in Anand, Gujarat, India, has been making significant contributions as a Research Associate in the Department of Agricultural Biotechnology at Anand Agricultural University, Gujarat, since 2022. Before this he served as a Senior Research Fellow at Cotton Research Station, Sardarkrushinagar Dantiwada Agricultural University, Talod, Gujarat, India for six months. He is expert in designing and interpreting the plant breeding and molecular breeding research. He is also expert in the high dimensional genome data analysis. He adeptly used the R and MS Excel including various other software like TASSEL, STRUCTURE, QTL cartographer, MapMaker, NTSys-pc, DARwin and Power Marker etc. to support his research. He is very much expert in variability analysis for this He and his coworkers developed the R package called "variability". He is also expert in handling multi-environment trait data and analysis. He is sound at various matting designs like L × T, Diallel mating design, GMA, and etc. His scholarly achievements extend to the publication of research papers in reputable journals, as well as the authorship of several book chapters and review papers that reflect his academic credit. Notably, he actively participates in the peer review process in several reputed journals. Beyond his academic responsibilities, Dr. Patel has served as an active plant breeder at Research Station for Distant Hybridization in Fruit and Field Crops at Department of Agricultural Biotechnology, Anand Agricultural University, Anand.

**Dr. Sushil Kumar** is an accomplished Assistant Professor in the Department of Agri. Biotechnology at Anand Agricultural University, Gujarat, since 2012. A proud alumnus of Rajasthan Agricultural University, Bikaner, he excels in Plant Molecular Biology and Molecular Breeding. Dr. Kumar has significantly contributed to the development of 12 varieties of various crops. His mentorship encompasses guiding 5 master's and one Ph.D. student. Dr. Sushil Kumar's academic achievements include over 100 publications, featuring in books from international publishers. The citations of his research papers exceed 1900 with h-index of 23, reflects his wide-reaching impact in the field of agricultural biotechnology. Additionally, he has led several externally and in-house projects funded by different agencies.

**Dr. Prity Kumari,** alumnus of Banaras Hindu University in Varanasi, Uttar Pradesh, India, has been making significant contributions as an Assistant Professor in the College of Horticulture at Anand Agricultural University, Gujarat, since 2015. Her research expertise spans a diverse range of areas, focusing on time series forecasting through statistical models and cutting-edge Deep Learning AI techniques. Dr. Kumari adeptly applies methodologies such as ARIMA, ARCH/GARCH, ANN, ML, Memory based Machine learning models and CNN to advance her research. Her scholarly achievements extend to the publication of research papers in reputable journals, as well as the authorship of several books and book chapters that reflect her academic credit. Notably, she actively mentors and guides numerous master's and Ph.D. students in the domain of agriculture. Beyond her academic responsibilities, Dr. Kumari has served as a visiting fellow at Western Sydney University, further enriching her research insights.

# Crop Trials with R

**Rumit Patel**
Research Associate
Department of Agri. Biotechnology
Anand Agricultural University
Gujarat, India

**Sushil Kumar**
Assistant Professor
Department of Agri. Biotechnology
Anand Agricultural University
Gujarat, India

**Prity Kumari**
Assistant Professor
College of Horticulture
Anand Agricultural University
Gujarat, India

**NIPA® GENX ELECTRONIC RESOURCES & SOLUTIONS P. LTD.**
New Delhi-110 034

**NIPA® GENX ELECTRONIC RESOURCES & SOLUTIONS P. LTD.**

101,103, Vikas Surya Plaza, CU Block
L.S.C.Market, Pitam Pura, New Delhi-110 034
Ph : +91 11 4386 0225, 9717133558, 9540816132
E-mail: newindiapublishingagency@gmail.com
Website: www.niparesources.com



Print ISBN: 978-81-19235-49-0

ebook ISBN: 978-93-58873-40-4



Composed and Designed by NIPA.

# Preface

In the ever-evolving landscape of statistical analysis and experimental design, the need for a resource that not only explains the fundamental concepts but also bridges the gap between theory and practical application is more pronounced than ever. This book, "Crop Trials with R" -a comprehensive guide to analysis of experimental design with R, is a response to that need. Our focus is to make the analysis of agricultural trail's/experiment's data accessible to everyone, including those who may be new to coding. By using the interface of R and providing screenshots, we make sure that even a non-statisticians can approach this powerful tool with confidence.

Our journey begins with an introduction to the basic concepts of statistics and R, the programming language that has become an indispensable tool for statisticians, researchers and data scientists across the globe. We explore why R is preferred for statistical computation and data visualization, supported by a user-friendly guide on its installation and setup. The core of this book is dedicated to experimental design, a critical aspect of scientific research that ensures the validity and reliability of results. We navigate through various designs like Completely Randomized Design (CRD), Randomized Block Design (RBD), and Latin Square Design (LSD), among others, explaining their principles, applications and nuances. Our approach is hands-on, with each chapter enriched by detailed examples and R code snippets. A significant portion of the book is focuses on the statistical methods of correlation and path analysis, emphasizing their types, applications and execution using R. As we progress, we emphasize the visualization of output, where we turn data into insights. This practical section of the book is designed to guide readers through multiple comparison tests and their visualization in R.

What sets "Crop Trials with R" apart is its focus on making statistical analysis in agriculture approachable and understandable. Whether you're a student, a professional or simply someone interested in agricultural data analysis, this book is designed to be your companion on a journey of learning, exploration, and discovery in the world of agricultural data analysis with R.

# Contents

# 1

# Introduction

This chapter describes fundamental terminology and lays the foundational framework for the course. It offers a general understanding of statistics, its purpose, and its role in addressing diverse problems across various fields of study. Additionally, it explains the rationale behind choosing R software as a powerful tool for data analysis

## Basic Concepts and Terminology

Statistics is an essential tool in agriculture, helping researchers to collect, analyze, interpret, and present numerical data related to plant traits, and agricultural practices. The fundamental goal is to make informed decisions and enhance crop yields, quality, and resilience.

To illustrate, researchers choose to work with a sample due to practical considerations, such as time and cost constraints. For instance, in a field with 10,000 crop plants, a researcher aims to determine the average plant height in the field. To conserve time and resources, they randomly select 200 plants from various locations within the field. These 200 plants constitute a sample from the entire population of 10,000 plants. By analyzing the height observed in the sample, the researcher can conclude the average height of the entire crop field.

Therefore, to make a scientific conclusion from sample to population, Statistics involved two fundamental branches:

A. Descriptive statistics.

B. Inferential statistics.

To grasp the distinction between descriptive and inferential statistics, it's essential to start with basic terminologies:

- Population consists of all the individuals or items of interest within a certain category.
    - Parameter is a numerical value which is derived from entire population.

- Sample represents a portion of the population and when chosen correctly, serves as a representative subset of the entire population.
    - Statistic is a numerical value derived from the data within a sample.
- Measurement is a numerical value or characteristic that is calculated for every individual within a population or a subset known as a sample
    - Qualitative data are measures of attributes or non-numerical characteristics.
    - Quantitative data are measurements of numerical characteristics.

Visual representation in the block diagram illustrates how statistics guides us in concluding experiments

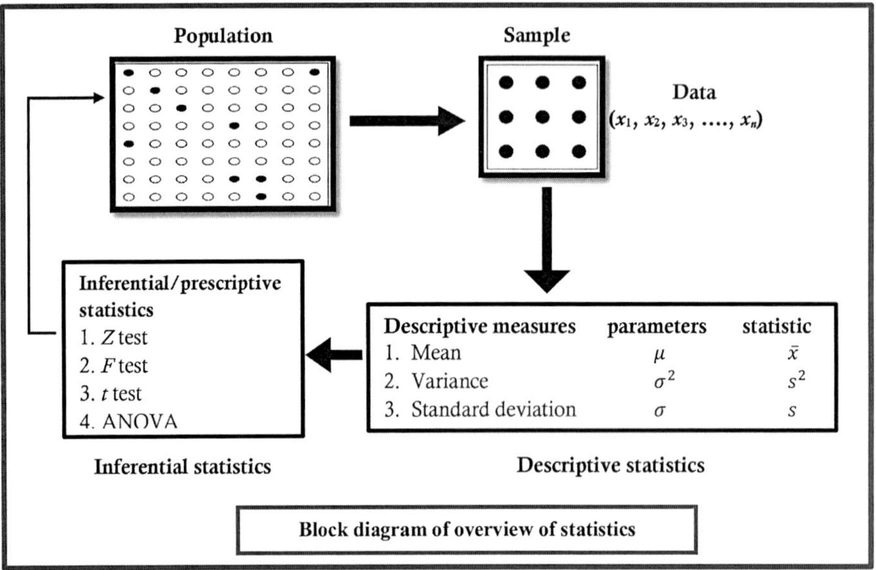

Block diagram of overview of statistics

## *Descriptive Statistics*

It involves the collection, organization, and presentation of numerical data collected from specific crops, plant varieties, or agricultural experiments. The aim is to characterize data sets, such as calculating measures of central tendency (e.g., mean, median) and measures of variability (e.g., standard deviation). For instance, a plant breeder may use it to calculate the average yield, disease resistance score, or nutrient content of a particular crop variety based on field trial data. It helps in the initial assessment of data.

### *Inferential Statistics*

Building on the foundation of descriptive statistics, inferential statistics examine hypothesis testing and concluding populations based on sample data. It is crucial in agriculture, particularly in cases where it's impractical to collect data from an entire population of plants. Instead, researchers gather data from a representative sample and use inferential statistics to conclude the entire population. Inferential statistics help breeders make predictions, assess genetic traits, and guide selection decisions with confidence.

For example: In a large-scale breeding program, a researcher wants to assess the disease resistance of a new tomato variety but cannot test every plant in the field. Instead, they take a random sample of tomato plants, evaluate disease resistance, and use inferential statistics to make predictions about the disease resistance of the entire crop.

## Function/Purpose of Statistics

In general, "Statistics" serves a dual role: it can refer to quantitative data (in plural form) used for empirical analysis and decision-making, while also representing a scientific discipline (in singular form) that underpins the collection and interpretation of such data.

- **a.** As a scientific discipline, statistics serves the following fundamental purposes
- **b.** Collection of data: It involves gathering information on traits like crop yield, disease resistance, and genetic markers. For instance, a plant breeder records the number of seeds produced by different hybrid tomato plants.
- **c.** Condensation: Convert data into meaningful summaries. An example is calculating the average height of multiple wheat plants from a breeding trial to assess overall performance.
- **d.** Estimation: Plant breeders often estimate traits like genetic heritability based on data from selected parent plants and their offspring. This estimation helps in predicting the traits of future generations.
- **e.** Tests of hypothesis: It helps in drawing inferences about a population from sample data. Suppose a new cross-breeding technique's effectiveness for enhancing drought tolerance in soybean plants is assessed through a series of experiments, and data collection on plant survival in drought conditions and employs statistical analysis to ascertain whether the new technique indeed proves effective in enhancing drought resilience.

## Limitations of Statistics

**a.** Data limitation: Statistics relies on available data and if the data is insufficient, incomplete, inaccurate, or biased, the results can be misleading.

**b.** Assumption dependence: Many statistical methods are based on specific assumptions and if these assumptions are violated, the results may not be valid.

**c.** Interpretation complexity: Statistical results can be challenging to interpret correctly, leading to potential misinterpretation.

**d.** Lack of causation: Statistics can establish correlations but not causation. Establishing cause-and-effect relationships often requires additional research and experimentation.

**e.** Sensitivity to outliers: Statistical analyses can be sensitive to outliers, which are extreme data points that differ significantly from the majority of the data.

## Why R Software

R is a powerful and versatile statistical programming language often preferred over licensed software tools such as SAS, SPSS, and MATLAB. This preference arises because, in research analysis, you must specify the use of licensed software due to its proprietary nature. In contrast, R and Python, both open-source options, offer distinct advantages over licensed software.

Key strengths of R over Python:

**a.** Specialized statistical packages: R has several specialized packages tailored for plant breeding and genetics analysis such as "lme4" for mixed-effects models, "pedigree" for pedigree analysis, and "qtl for quantitative trait locus analysis. These packages are well-established and widely used in the field.

**b.** Bioconductor: R hosts the Bioconductor project, a collection of over 2500 packages specifically designed for bioinformatics and computational biology tasks. This includes packages for genomics, transcriptomics, and other plant-related data analyses.

**c.** Data visualization in plant science: R's ggplot2 package is renowned for its expressive data visualization capabilities. "ggplot2" is based on the "Grammar of Graphics" concept, which provides a structured and intuitive approach to creating visualizations.

R's ggplot2, along with specialized packages like "ggbio," can be particularly useful for creating customized and informative plots for visualizing genetic data, gene expression, and plant phenotypes.

### *Limitations of R over Python*

**a.** General-purpose vs. statistical language: Python is a general-purpose programming language with a wide range of libraries, including powerful statistical and data analysis libraries like NumPy, pandas, and sci-kit-learn. While R excels in statistical analysis, Python's versatility makes it more suitable for end-to-end data science projects that involve web development, automation, and other tasks beyond statistics.

**b.** Learning curve for non-statisticians: R's syntax and data manipulation techniques can be challenging for individuals with a non-statistical background. Python, on the other hand, has a more straightforward syntax that can be easier to learn for those who are new to programming and data analysis.

**c.** Performance and scalability: For large-scale data analysis and complex machine learning tasks, Python may have an advantage due to its performance optimization libraries like NumPy and integration with tools like TensorFlow and PyTorch. R may face limitations in terms of scalability and performance for extremely large datasets.

# 2

# Installation of R

This chapter serves as an essential primer on R, providing step-by-step instructions for installing both R and RStudio, one of the most popular integrated development environments (IDEs) for R. Additionally, the chapter covers important aspects such as importing data into R and setting the working directory.

## Install R and RStudio

R and RStudio can be installed on Windows, MAC OSX, and Linux platforms. RStudio (the IDE of R) offers a code editor, a unified console, and a suite of visualization tools. It simplifies the use of R, making it more accessible and efficient for data analysis and statistical computing tasks.

Steps to install R and RStudio

1. R can be downloaded and installed from the Comprehensive R Archive Network (CRAN) webpage (http://cran.r-project.org/).
2. After installing R software, install the RStudio software available at: http://www.rstudio.com/products/RStudio/.
3. Launch RStudio and start using R inside R studio.

## Appearance of RStudio

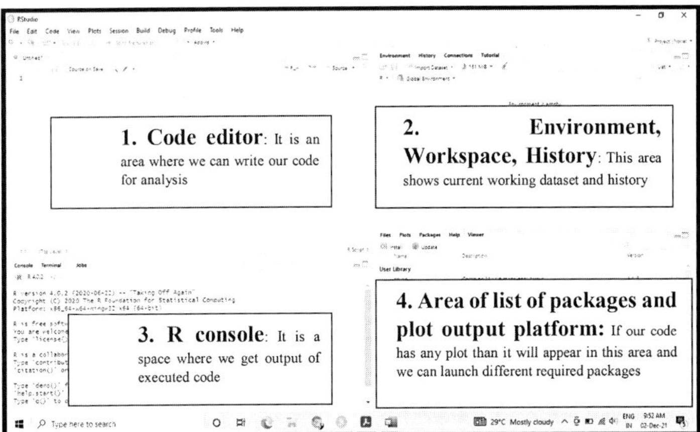

## How to Download Any Package?

CRAN, Bioconductor (biology-related R packages), and GitHub repositories all have thousands of R packages accessible for download and installation.

How do I install CRAN packages?

There are two methods to install any package

*Method 1*: Using **install.packages** Function

1. Write **install.packages** ("variability") in the code editor, where "variability" is the name of the package you want to install
2. Press Ctrl+Enter or click on Run the button in the code editor area
3. It will start downloading the variability package from CRAN

✓ Note: Ensure that your PC is connected to the internet

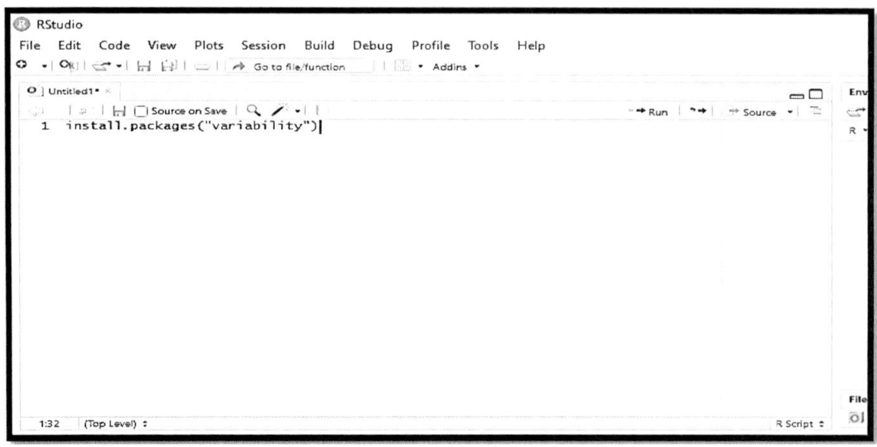


*Method 2*: Using the **Package Manager**

1. Locate the "**Packages**" section in your **R environment**
2. In this area, find and select the package you want to install
3. Look for the "Install" option associated with the selected package
4. Click on "**Install**"

- This will open a dialog box (Install Packages)

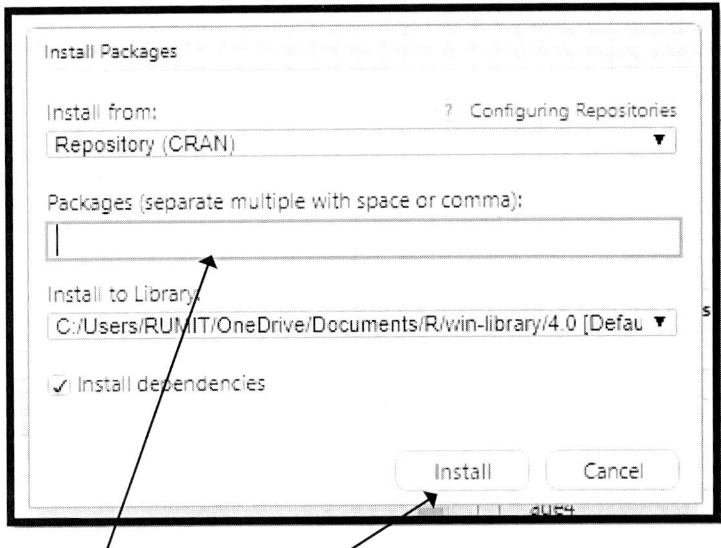


- Write package name and click on install
- This will start the downloading of the package from CRAN with the required dependencies
- Now, you can find your package from the list of packages

- If it does not appear to be a downloaded package, then press the **refresh** button given in the 4th quadrant

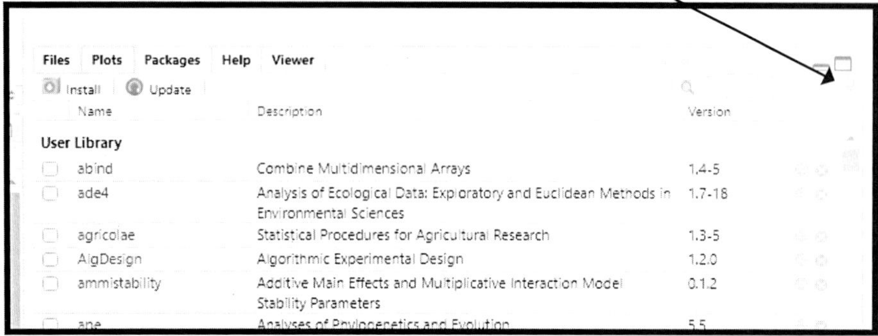


How to launch a package for data analysis?

There are two different ways:

1. First, load the package to use the functions in the package
2. Write (*in small letters*) the library (for example, "**variability**") in the first quadrant or script editor like shown below
3. Press Ctrl+Enter or click on run

**Note:** before pressing **Ctrl+Enter** or pressing on run button, ensure that cursor in the script editor must be within your written code like shown in pic

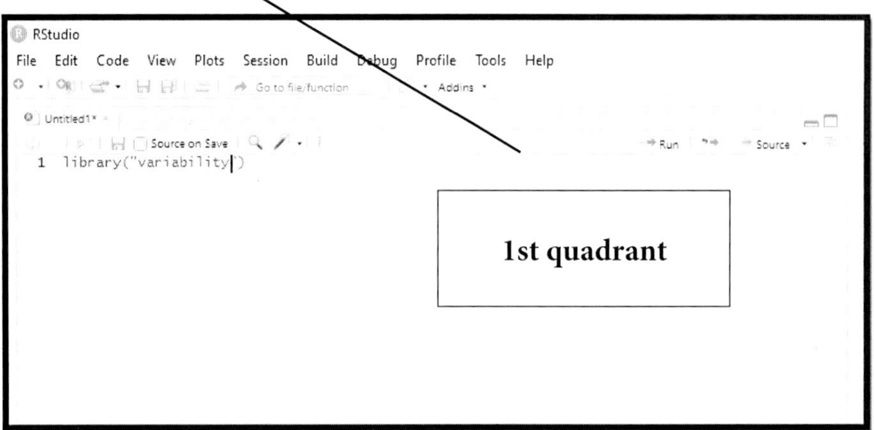

The expected output of library("variability"): You can find it in console (3rd quadrant) and the package area (4th quadrant).

**Console (3rd Quadrant)**

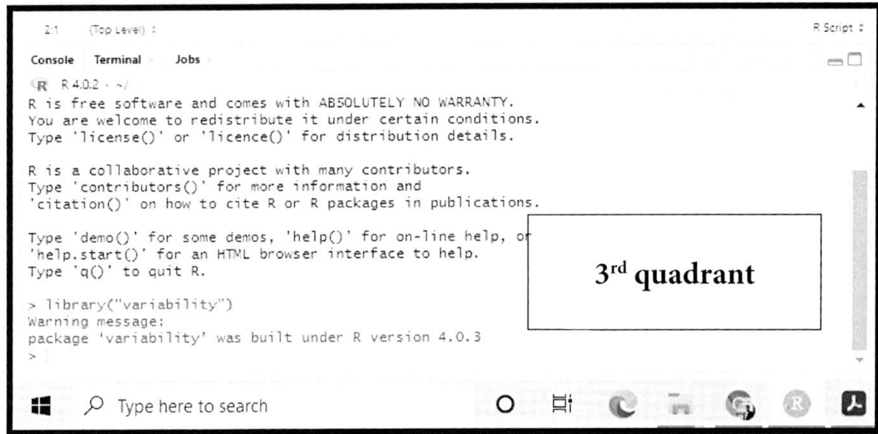


**Package Area (4th Quadrant)**

- ✓ Type the name of the package, such as "variability," in the search bar within the package management area
- ✓ Locate the package in the search results and check the square box next to it.

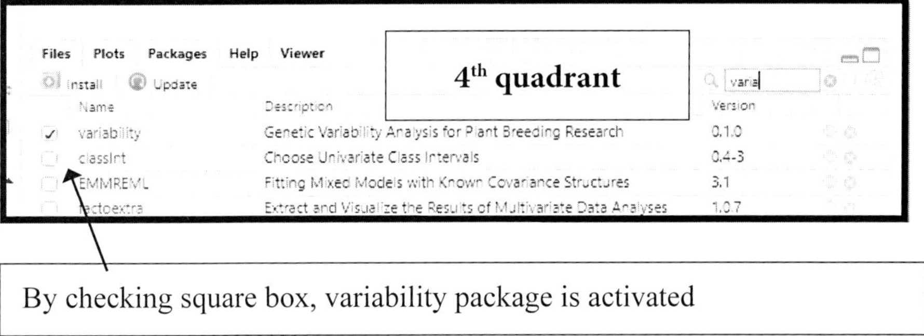


By checking square box, variability package is activated

## How to Import Input Data in R?

Data can be in the .xlx/.xlsx, .csv, and .txt format. Each format uses different code in R. The best format of data input in R is .xlx and .csv

1. *Import Data in R by Point and Click Option: .xlsx Format*
   - Click on the "**Import Dataset**" button or option

- In the dialog box that appears, select "From Excel" as your data source
- A file dialog will open. Click on the "Browse" button to navigate the location of your .xlsx file in your computer's folders
- Once you've located your .xlsx file, select it and click on the "Import" button
- Your dataset will be imported into RStudio, ready for analysis

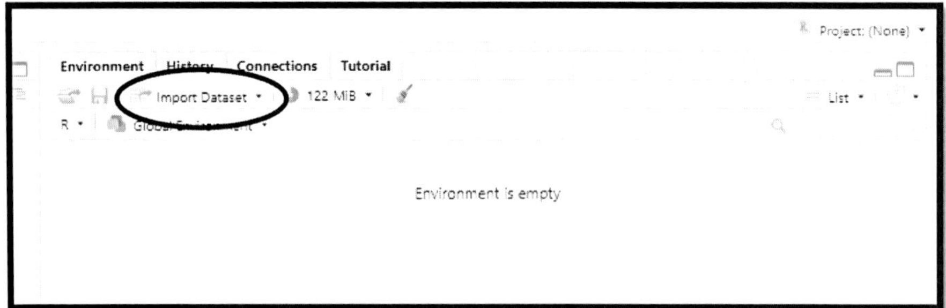


1. Click on Import Dataset button

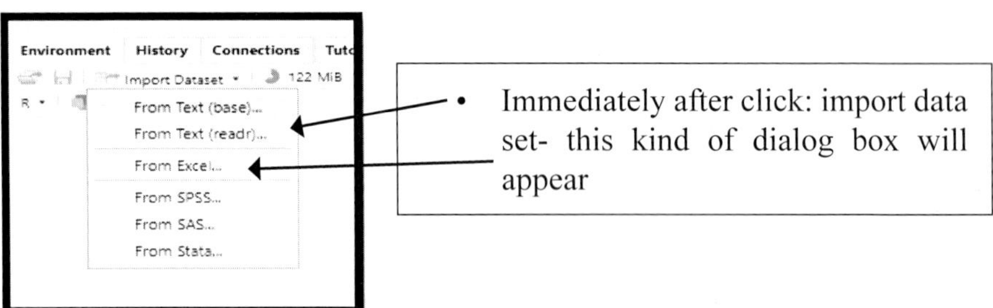


After selecting "**From Excel**" a dialog box will appear

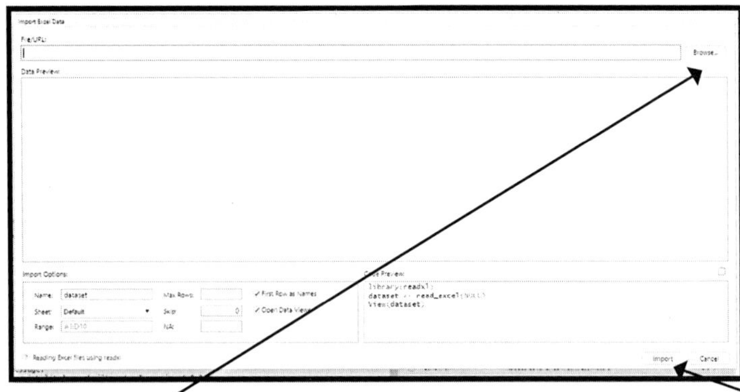

Now, click on browse and choose your datasheet from your folder and click on import

Your dataset is inserted in RStudio

Description of the above dialogue box

This is the address of our datasheet

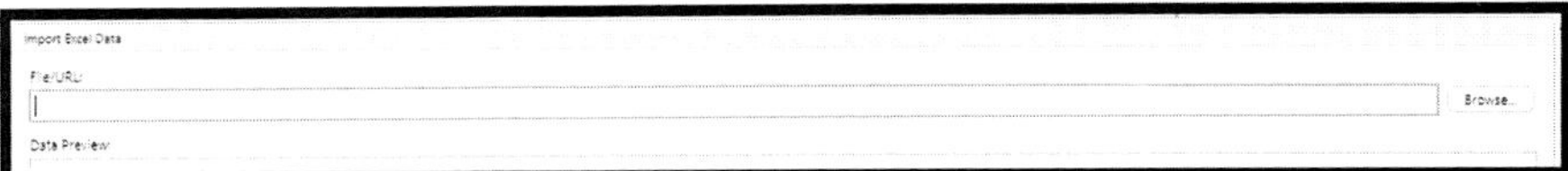


This is an important option box

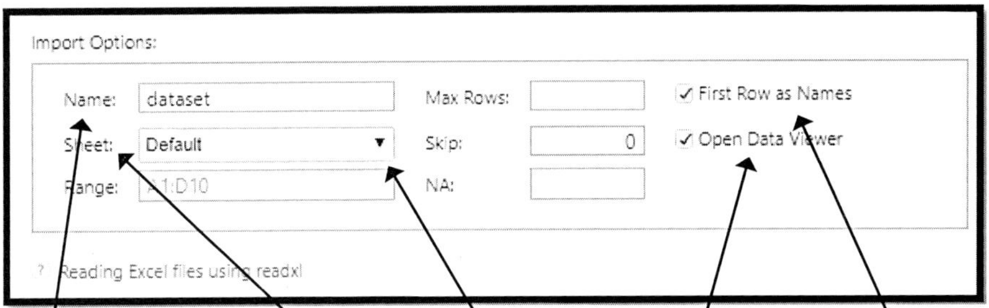

Import Options:

Name: dataset | Max Rows: | ✓ First Row as Names

Sheet: Default | Skip: 0 | ✓ Open Data Viewer

Range: A1:D10 | NA:

? Reading Excel files using readxl

**Name** indicates the name of our data which will be used for analysis. We can change name of our data as per our requirement/choice but make sure that final name should always be used during analysis

**Sheet** indicates the name of sheet from our data will be imported. If our data in sheet 2 or in another sheet in same file than we can change the sheet address by using its **dropdown menu**

These two options must be activated if our data set has header and you wish to see data after import.

**Once you click on the "import" button—RStudio will look like this**

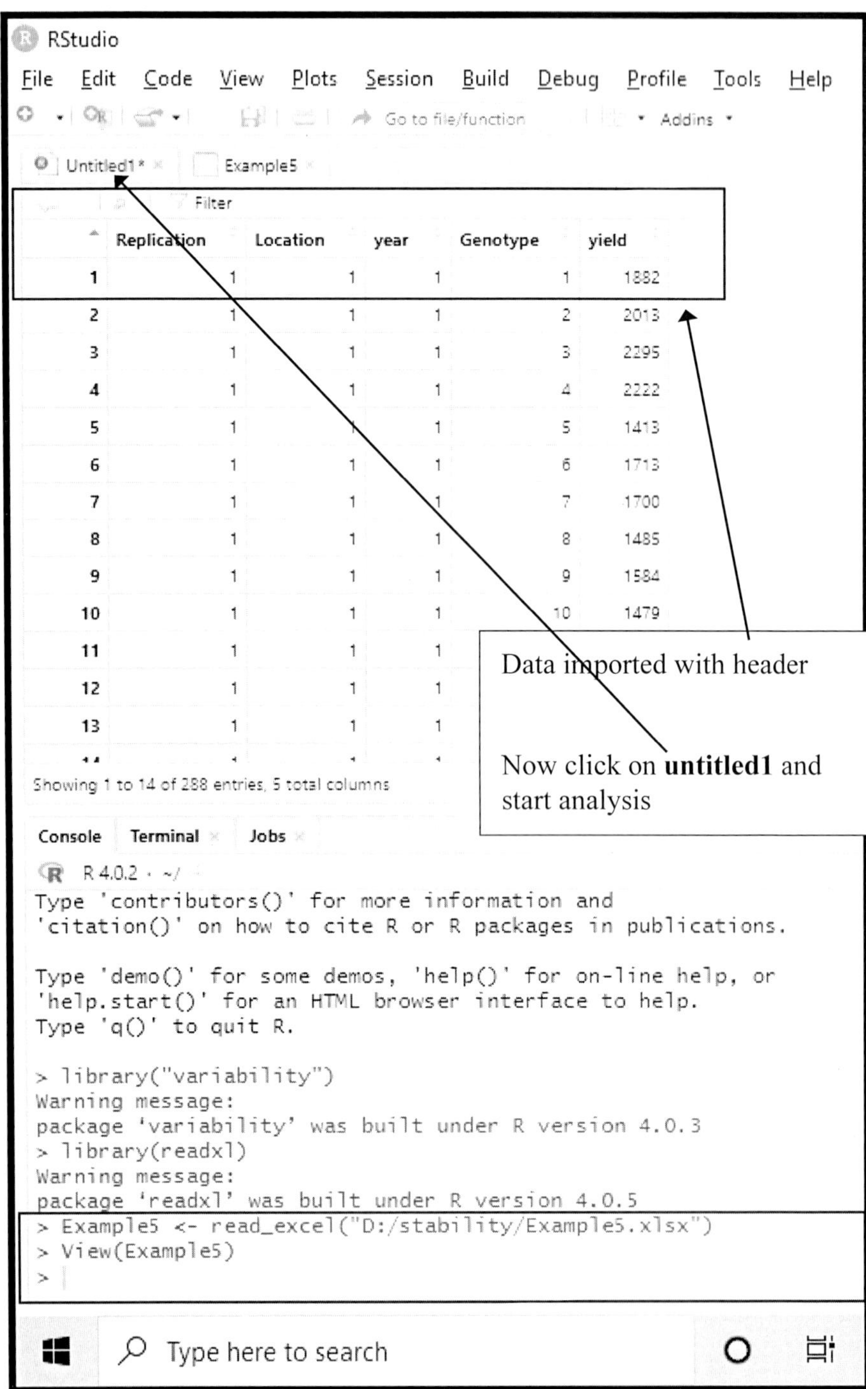
RStudio
File Edit Code View Plots Session Build Debug Profile Tools Help
Go to file/function
Addins
Untitled1*
Example5
Filter
Replication Location year Genotype yield
1 1 1 1 1 1882
2 1 1 1 2 2013
3 1 1 1 3 2295
4 1 1 1 4 2222
5 1 1 1 5 1413
6 1 1 1 6 1713
7 1 1 1 7 1700
8 1 1 1 8 1485
9 1 1 1 9 1584
10 1 1 1 10 1479
11 1 1 1
12 1 1 1
13 1 1 1
Showing 1 to 14 of 288 entries, 5 total columns
Data imported with header
Now click on **untitled1** and start analysis
Console Terminal Jobs
R 4.0.2 · ~/
Type 'contributors()' for more information and
'citation()' on how to cite R or R packages in publications.
Type 'demo()' for some demos, 'help()' for on-line help, or
'help.start()' for an HTML browser interface to help.
Type 'q()' to quit R.
> library("variability")
Warning message:
package 'variability' was built under R version 4.0.3
> library(readxl)
Warning message:
package 'readxl' was built under R version 4.0.5
> Example5 <- read_excel("D:/stability/Example5.xlsx")
> View(Example5)
>
Type here to search

2. *Import Data in R by coding: .csv Format*

   Write the following code in your script:

   Data=read.csv ("File path with file name and extension") then press Ctrl+Enter.

✓ Note: If you copy the file path and paste it into your script, ensure that you replace any **backslash (\)** in the file path with **forwarding Slash (/)**

```
library("variability")

Data= read.csv("C:/Users/RUMIT/OneDrive/Desktop/RCBDdata.csv")
```

**Result in the Console**

```
> Data= read.csv("C:/Users/RUMIT/OneDrive/Desktop/RCBDdata.csv")
>
```

You can see that data is imported in the **Data** folder.

Similarly, if data is in .txt format, use the following code:

Data = read.txt ("file path along with file name with .txt extension") and execute the code in R.

✓ **Note:** .csv and .txt formats are commonly used for data import in R due to their simplicity and compatibility with the language.

## Setting and Checking the Working Directory

The working directory in R defines the folder location where you are currently operating. Without setting a working directory, you would need to provide the full path to your files for reading and analysis, which can be cumbersome. Setting the working directory simplifies this process, as it tells R to look for files in that specific folder and save results there.

### *How to Set a Working Directory?*

In the script editor, use the following code:

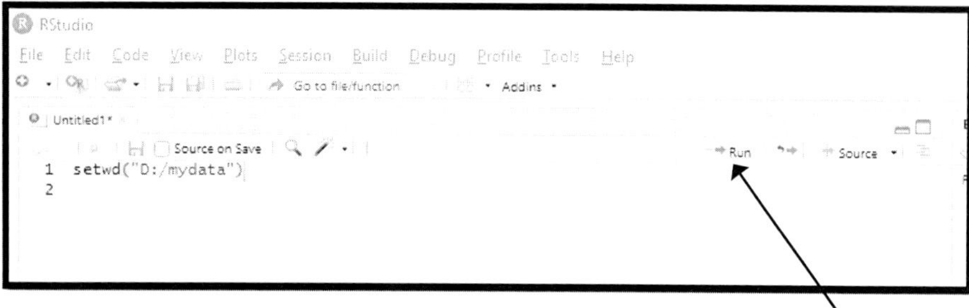


✓ Replace "Your Directory Path" with the path to the folder where your files are located

✓ Place the cursor within this code and press Ctrl+Enter or click "Run" and you will find the result in the console like below:

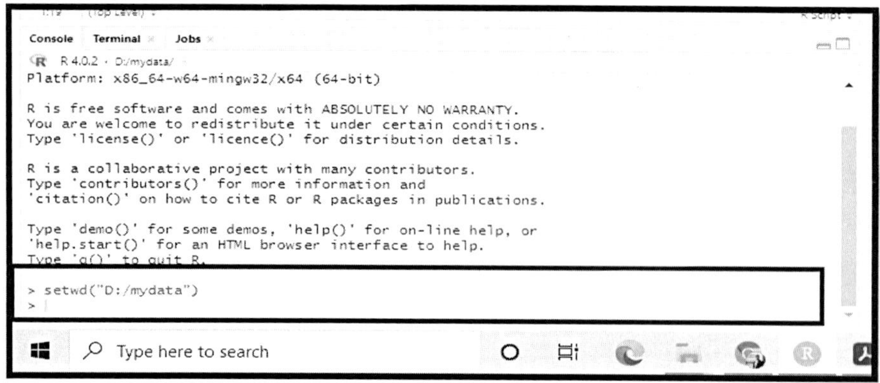


**Important note:**

When you copy a file path from your computer, it may look like this: "D:\mydata." However, when you paste this link into R, it won't work as expected because **R uses the "/" symbol to read directories**. To ensure proper functionality, you need to **replace the "\" with "/"** in the file path when pasting it into R.

### *Checking the Current Working Directory*

To confirm the current working directory, you can use the following code:

getwd()    #for getting the path of the current directory

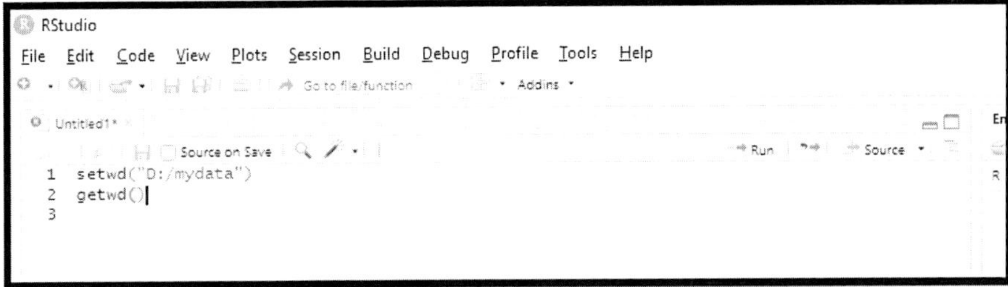


Press Ctrl+Enter

**Find the result in a console**

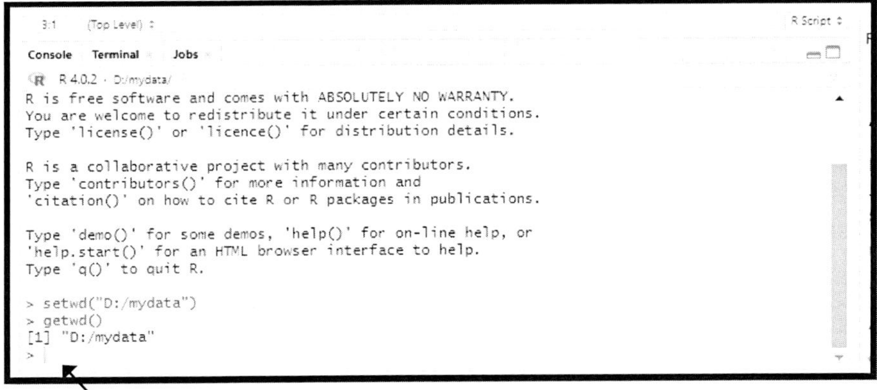


It suggests the current working directory!!!

We are currently working in D:/mydata (means D drive contain folder named "mydata"

# 3

# Basic Statistics

In this chapter, the essentials of basic statistics are explored, with a focus on obtaining all key measures of basic statistics with just a single click. Additionally, the process of creating informative data distribution graphs is also explained.

## Basic Statistics

Basic statistics offer fundamental tools that help us to make sense of data. Below, you'll find concise descriptions of key basic statistical measures.

i. **Arithematic mean (average)**

   The average represents the central value of a dataset, calculated by summing all values and dividing by the number of data points.

ii. **Median (middle value)**

   It is the middle value in a dataset when all values are arranged in ascending order. It's less sensitive to outliers.

iii. **Mode (most common value)**

   The mode is the value that appears most frequently in a dataset.

iv. **Variance**

   Variance measures how much individual data points deviate from the mean. A higher variance indicates a greater spread.

v. **Standard deviation**

   The standard deviation is the square root of the variance. It provides a measure of how spread out the data is.

vi. **Range**

   It is the difference between the maximum and minimum values in a dataset.

vii. **Skewness**

Skewness can be quantified as representing the extent to which a given distribution varies from a normal distribution. A normal distribution has a skew of zero. It measures the asymmetry of the data distribution. Positive skew indicates a longer tail on the right, while negative skew has a longer left tail.

viii. **Kurtosis**

It quantifies the tailedness or peakedness of a data distribution. High kurtosis indicates heavy tails, while low kurtosis suggests light tails. It differs from the tails of a normal distribution. In other words, kurtosis identifies whether the tails of a given distribution contain extreme values

## Basic Statistics in a Single Line Code

**Required package: metan** ##if not installed first install it

```
# Load the "metan" package
library(metan)
# Perform a comprehensive basic statistics analysis
q <- desc_stat(
 x,                    # Your dataset
 stats = "all",        # Request all available statistics
 hist = TRUE,          # Create a histogram
 level = 0.95,         # Set the confidence level to 95%
 digits = 4,           # Specify the number of decimal places
 na.rm = FALSE,  # Removes missing values (if TRUE)
 verbose = TRUE,              # Display details of the analysis
 plot_theme = theme_metan()            # Apply a specific plot theme
)
```

**Write the code in the script editing area like below**

```
# Load the "metan" package
library(metan)

# Perform a comprehensive basic statistics analysis
q <- desc_stat(
  x,                  # Your dataset
  stats = "all",      # Request all available statistics
  hist = TRUE,        # Create a histogram
  level = 0.95,       # Set the confidence level to 95%
  digits = 4,         # Specify the number of decimal places
  na.rm = FALSE,      # Remove missing values (if TRUE)
  verbose = TRUE,     # Display details of the analysis
  plot_theme = theme_metan()  # Apply a specific plot theme
)
```

✓ Note: that the statistic names are not case-sensitive. Both commas or space can be used as a separator.

## *Description of Code*

- **q:** This variable stores the subset where the analysis results will be saved
- **desc_stat:** This is the name of the function available from the "metan" package, used for calculating descriptive statistics
- **x:** This is the dataset or subset where your data is stored, and you want to perform the analysis
- **stats:** This parameter allows you to specify which statistics you want to compute. To select multiple statistics, use names like stats = c("median", "mean", "cv", "n"). You can choose from a variety of options, such as:
  - "n": the length of the data
  - "n.valid": The valid (Not NA) number of elements
  - "n.missing": The number of missing values
  - "n.unique": The length of unique elements
  - "ps": The pseudo-sigma (iqr/1.35)
  - "q2.5", "q25", "q75", "q97.5": Percentile 2.5\quartile, third quartile, and percentile 97.5
  - range: The range of data
  - "sd.amo", "sd.pop": The sample and population standard deviation

- "se": The standard error of the mean
- "skew": skewness
- "sum": The sum of the values
- "sum.dev": The sum of the absolute deviations
- "sum.sq.dev": The sum of the squared deviations
- "n.valid": The size of the sample with a valid number (not NA)
- "var.amo", "var.pop": The sample and population variance

- **hist:** When set to TRUE, this parameter creates a histogram for each selected variable
- **level:** You can specify the confidence level (default is 0.95) for calculating the confidence interval of the mean
- **digits:** This parameter determines the number of significant digits in the results
- **na.rm:** A logical argument. If set to TRUE, it removes missing values from the dataset during analysis (default is FALSE)
- **verbose:** Another logical argument. When set to FALSE, the code runs without displaying detailed information
- **plot_theme:** This parameter defines the graphical theme for any plots generated as part of the analysis. The default theme is theme_metan()

## *Output in Console*

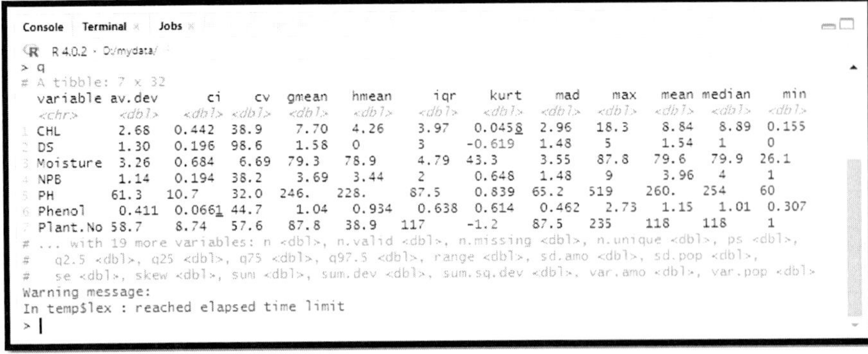

```
Console   Terminal   Jobs
R 4.0.2 · D:/mydata/
> q
# A tibble: 7 x 32
  variable av.dev      ci     cv  gmean  hmean    iqr    kurt    mad    max   mean median    min
  <chr>     <dbl>   <dbl>  <dbl>  <dbl>  <dbl>  <dbl>   <dbl>  <dbl>  <dbl>  <dbl>  <dbl>  <dbl>
1 CHL       2.68   0.442   38.9    7.70   4.26   3.97   0.0458  2.96  18.3    8.84   8.89  0.155
2 DS        1.30   0.196   98.6    1.58   0      3     -0.619   1.48   5      1.54   1     0
3 Moisture  3.26   0.684    6.69  79.3   78.9    4.79  43.3     3.55  87.8   79.6   79.9   26.1
4 NPB       1.14   0.194   38.2    3.69   3.44   2      0.648   1.48   9      3.96   4      1
5 PH       61.3   10.7     32.0  246.   228.    87.5    0.839  65.2  519    260.   254     60
6 Phenol     0.411  0.0661 44.7    1.04   0.934  0.638  0.614   0.462  2.73   1.15   1.01  0.307
7 Plant.No 58.7    8.74    57.6   87.8   38.9  117     -1.2    87.5  235    118    118      1
# ... with 19 more variables: n <dbl>, n.valid <dbl>, n.missing <dbl>, n.unique <dbl>, ps <dbl>,
#   q2.5 <dbl>, q25 <dbl>, q75 <dbl>, q97.5 <dbl>, range <dbl>, sd.amo <dbl>, sd.pop <dbl>,
#   se <dbl>, skew <dbl>, sum <dbl>, sum.dev <dbl>, sum.sq.dev <dbl>, var.amo <dbl>, var.pop <dbl>
Warning message:
In temp$lex : reached elapsed time limit
> |
```

After running your "metan" analysis, the results are stored in a subset called "q" but you might not see the complete results on your console. To access and save these results in .text file, follow these steps:

The results of "metan" does not sink usually in a text file. The results are stored in a *tibble* format. Therefore, to extract the result from *tibble*, write the below-given code in the script editor.

```
15
16  sink("basic stat.txt")
17  print(q, n = 6, width = Inf)
18  sink()
19
```

**Line 16:** This code creates a text file named "basic stat.txt" in the working directory. It's where we'll save our statistical results.

**Line 17 (description of code)**

> **print(*q*, *n* = 6, width = Inf)** is used to display and print the results of our analysis in the RStudio console.
>
> ***q*** refers to the name of the subset where our analysis results are stored.
>
> ***n* = 6** specifies that we want to print only the first six rows of the result. In this context, it means we're interested in the statistics of the first six variables but not the seventh variable, which appears to be related to "plant no."
>
> **width = Inf** ensures that all columns containing the results are displayed without any truncation.

**Line 18:** This line completes the process of writing the results to the "basic stat.txt" text file. Essentially, it saves the displayed results in the console to the specified text file.

**Saved text file**

In the resulting text file, you might see the message "# … with 1 more row" because we've chosen to display only the first six rows of our analysis, so the result of 7th line is hidden

```
basic stat - Notepad
File  Edit  Format  View  Help
# A tibble: 7 x 32
  variable av.dev      ci    cv  gmean   hmean    iqr     kurt    mad    max   mean median    min
  <chr>     <dbl>   <dbl> <dbl>  <dbl>   <dbl>  <dbl>    <dbl>  <dbl>  <dbl>  <dbl>  <dbl>  <dbl>
1 CHL        2.68  0.442  38.9    7.70   4.26   3.97   0.0458  2.96   18.3    8.84   8.89  0.155
2 DS         1.30  0.196  98.6    1.58   0      3     -0.619   1.48    5      1.54   1     0
3 Moisture  3.26   0.684   6.69  79.3   78.9    4.79  43.3     3.55   87.8   79.6   79.9  26.1
4 NPB        1.14  0.194  38.2    3.69   3.44   2      0.648   1.48    9      3.96   4     1
5 PH        61.3  10.7    32.0  246.   228.    87.5    0.839  65.2   519    260.   254    60
6 Phenol     0.411 0.0661 44.7    1.04   0.934  0.638  0.614   0.462  2.73   1.15   1.01  0.307
      n n.valid n.missing n.unique     ps   q2.5     q25    q75  q97.5  range sd.amo sd.pop     se
  <dbl>   <dbl>     <dbl>    <dbl>  <dbl>  <dbl>   <dbl>  <dbl>  <dbl>  <dbl>  <dbl>  <dbl>  <dbl>
1   235     235         0      235  2.94   1.71    6.80   10.8   15.2   18.1   3.44   3.43  0.224
2   235     235         0        6  2.22   0       0       3      5      5     1.52   1.52  0.0993
3   235     235         0      233  3.55  71.0    77.6    82.4   86.3   61.7   5.32   5.31  0.347
4   235     235         0        9  1.48   2       3       5      8      8     1.51   1.51  0.0987
5   235     235         0      130 64.8   98.5   210.    298    436.   459    83.4   83.2   5.44
6   235     235         0      214  0.473  0.364   0.781  1.42   2.49   2.42  0.514  0.513 0.0335
    skew    sum sum.dev sum.sq.dev var.amo var.pop
   <dbl>  <dbl>   <dbl>      <dbl>   <dbl>   <dbl>
1 -0.115  2076.    630.      2770.   11.8    11.8
2  0.699   363     306.       542.    2.32    2.31
3 -4.67  18699.    767.      6632.   28.3    28.2
4  0.954   930     267.       536.    2.29    2.28
5  0.433 61177   14394.  1628516.  6959.   6930.
6  0.906   270.     96.6      61.9    0.264   0.263
# ... with 1 more row
```

It also generates a histogram if we activate the "hist" (histogram) function in the code as described above.

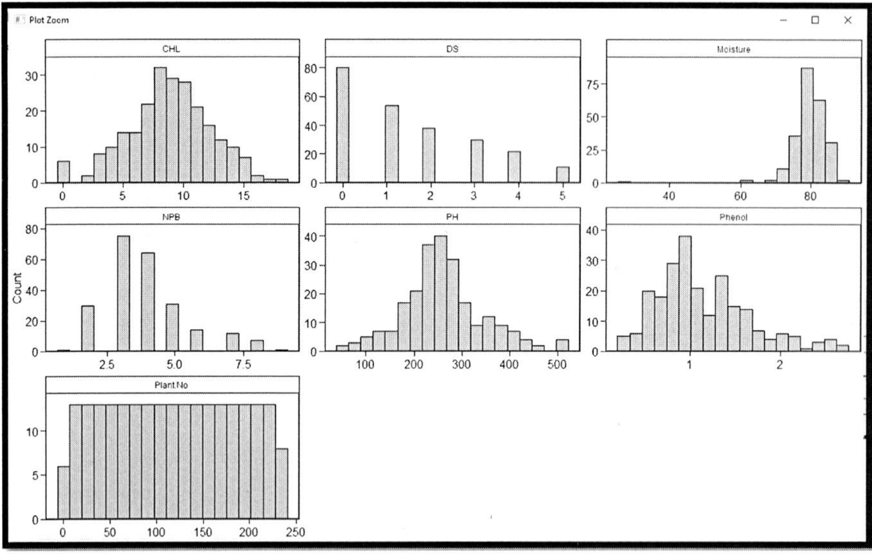

## Analyzing Data Distribution and Visualizing Results

If you want to know the distribution of data and calculate skewness kurtosis expertly then:

### Skewness

**Package required:** To calculate skewness, you need the "moments" package.

**Code:** skewness()

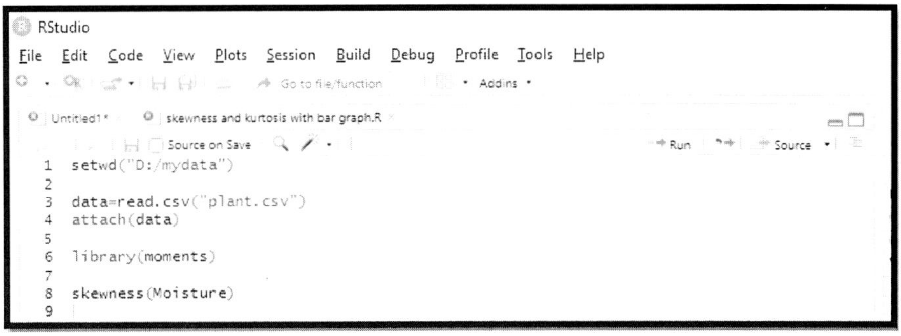

```
setwd("D:/mydata")

data=read.csv("plant.csv")
attach(data)

library(moments)

skewness(Moisture)
```

### Result in console

```
> library(moments)
> skewness(Moisture)
[1] -4.635573
>
```

### Kurtosis

**Package required:** The "moments" package is also required for kurtosis calculations.

**Code:** kurtosis()

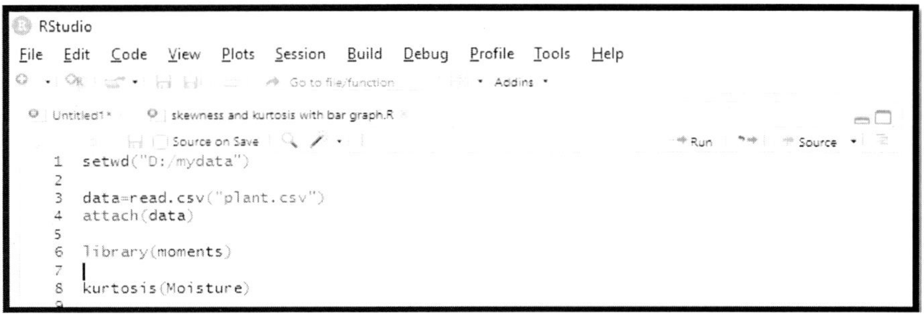

```
setwd("D:/mydata")

data=read.csv("plant.csv")
attach(data)

library(moments)

kurtosis(Moisture)
```

## Result in console

### *Histogram with Population Distribution Line*

**Package required:** Creating a histogram does not require any additional package. The “hist” function is inbuilt into R.

**Code:** To draw a histogram with a population distribution line, you can use the “hist” function.

**Result in plot area:** This code generates a histogram with the population distribution line.

```
Console   Terminal   Jobs
R 4.0.2 · D:/mydata/
> library(moments)
> kurtosis(Moisture)
[1] 45.38419
>
```

```
RStudio
File Edit Code View Plots Session Build Debug Profile Tools Help
Untitled1*   skewness and kurtosis with bar graph.R
Source on Save                                   Run   Source

setwd("D:/mydata")

data=read.csv("plant.csv")
attach(data)

h<-hist(Moisture, breaks=10, col="light blue", xlab="Plant Height in F2 generation",
        main="Distribution for Plant Height")

xfit<-seq(min(Moisture),max(Moisture),length=100)
yfit<-dnorm(xfit,mean=mean(Moisture),sd=sd(Moisture))
yfit <- yfit*diff(h$mids[1:2])*length(Moisture)
lines(xfit, yfit, col="black", lwd=2)
```

## Result in plot area

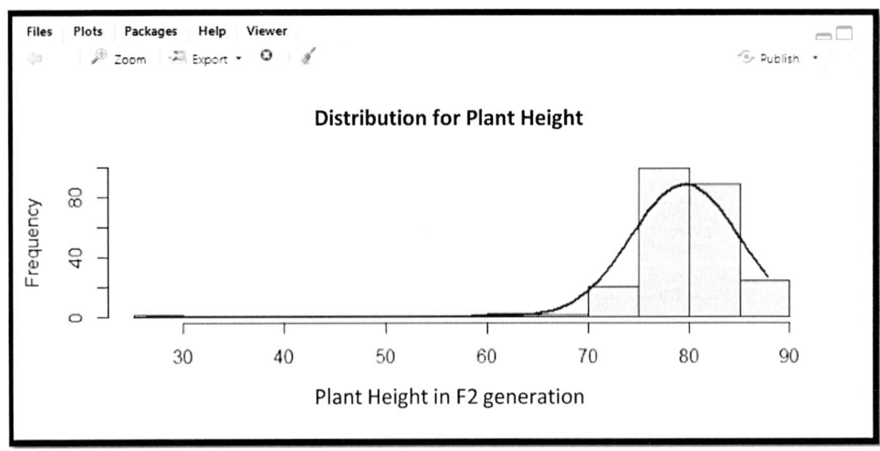

# 4

# Experimental Design with R

In this chapter, the focus is on understanding the importance of experimental design and being familiar with key terminology. You'll also learn about the three fundamental principles of experimental design and their roles in conducting experiments. Additionally, you'll learn how to do randomization of treatments in various experimental designs using R.

## Experimental Design

Experimental design is the methodical planning of scientific experiments for reliable results by controlling variables and minimizing biases. It serves as the foundation for effective research. Some key definitions to enhance understanding of this concept:

Key terminologies

- Experiment: Structured investigation for knowledge acquisition or validation.
- Experimental unit: Entity receiving treatment (e.g., land plot, pot).
- Treatment: Specific procedure for assessment and comparison.
- Experimental error: Variation among identically treated units.

## Principles of Experimental Design

Experiments are vital tools in agricultural research, aiming to assess treatment effectiveness and quantify their impact. However, extraneous factors can confound results. To mitigate these issues, we depend on three core principles of experimental design: replication, randomization, and local control.

### *Replication*

Repeating treatments in an experiment estimate error, enhance precision, and maintain error control.

### *Randomization*

Randomly assigning treatments ensures unbiased estimates and supports valid significance testing.

### *Local control*

By reducing systematic influences and promoting homogeneity among nearby units, local control minimizes experimental error in various plot arrangements. Understanding and applying these principles enhances research reliability. When a treatment appears more than.

Out of the three principles, the experimenter decides the number of repetitions and local control based on their experiment's needs, requiring no software assistance. However, the allocation of treatments to experimental units, a critical aspect, is facilitated using R. In the following section, we will explore the process of random treatment allocation in completely randomized design (CRD), randomized block design (RBD), Latin square design (LSD) designs, split-plot design, and strip plot design, offering comprehensive guidance.

✓ Note: The explanation of CRD, RBD, and LSD concepts will be provided in later chapters.

Randomization can be done through many methods using RStudio and R software. Before starting any analysis, you must set the working directory (explained in Chapter 2). It requires the package **"agricolae".**

### Randomization for CRD

i. Begin by defining treatment names in a .csv file as shown below:

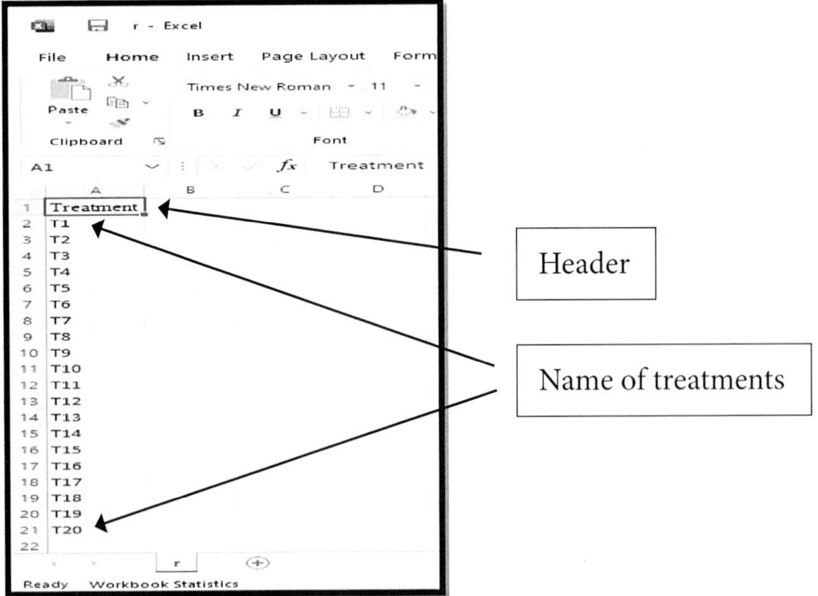

ii. Read this file in RStudio

iii. Type: treatment = read.csv("treatments.csv")

This will look like this in the console.

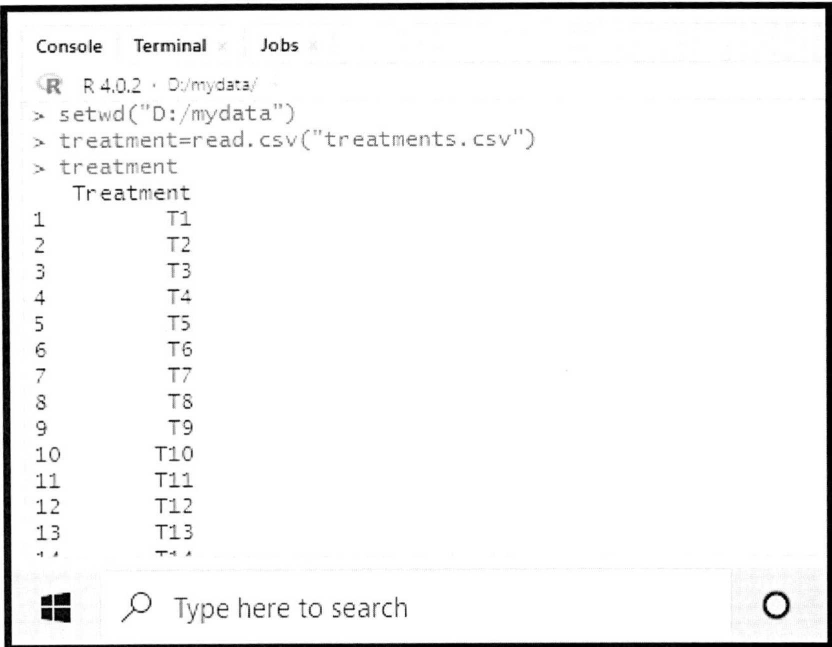


To randomize the treatments listed in the file, follow these steps:

iv. Attach file by typing code: attach()

v. In your script editor, enter the following code:

design.crd(treatment, r = 2, serie = 1, seed = 0, kinds = "Wichmann Hill," randomization = TRUE)

RStudio

File Edit Code View Plots Session Build Debug Profile Tools Help

Untitled1*

```
library("agricolae")
setwd("D:/mydata")

treatment=read.csv("treatments.csv")
attach(treatment)

design.crd(treatment, r=2, serie = 1, seed = 0, kinds = "Wichmann-Hill",randomization=TRUE)
```

**Run** and see the result in the console

```
Console  Terminal  Jobs
R 4.0.2 · D:/mydata/
> design.crd(treatment, r=2, serie = 1, seed = 0, kinds = "Wichmann-Hill",randomization=TRUE)
$parameters
$parameters$design
[1] "crd"

$parameters$trt
   Treatment
1         T1
2         T2
3         T3
4         T4
5         T5
6         T6
7         T7
8         T8
9         T9
10       T10
```

## Understanding the arguments in the code

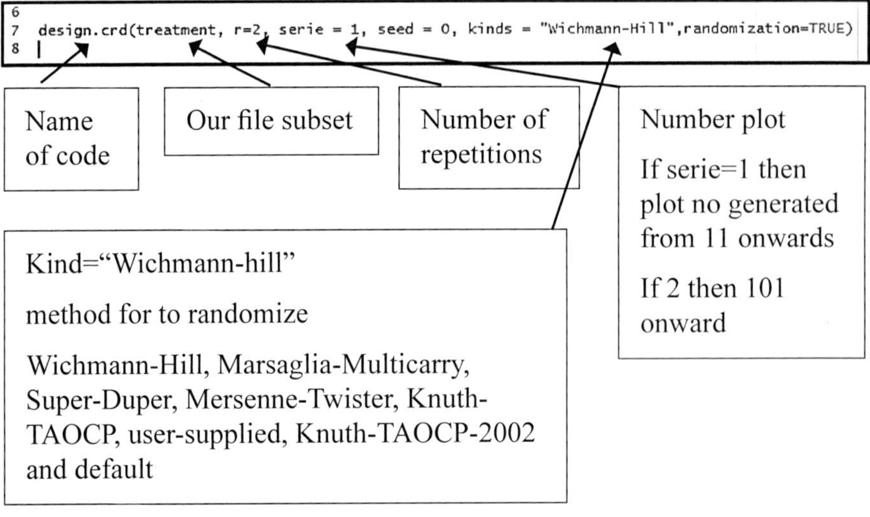


## Run the "design.crd" with a subset like below

```
treat=design.crd(treatment, r=2, serie = 1, kinds = "Wichmann-Hill",randomization=TRUE)
```

To extract the randomized treatments, which are stored under the name "book" and save them in a text file, you can use the following code in your script editor:

write.table(treat$book, "crd.txt," row.names = F, sep="\t")

```
RStudio
File Edit Code View Plots Session Build Debug Profile Tools Help

library("agricolae")
setwd("D:/mydata")

treatment=read.csv("treatments.csv")
attach(treatment)

treat=design.crd(treatment, r=2, serie = 1, kinds = "Wichmann-Hill",randomization=TRUE)

write.table(treat$book,"crd.txt", row.names = F, sep="\t")
```

To view the resulting text file, you can do one of the following:

**a.** Navigate to the "mydata" folder (if specified) or check your working directory.

**b.** You can also use the following code to open the file directly from R:

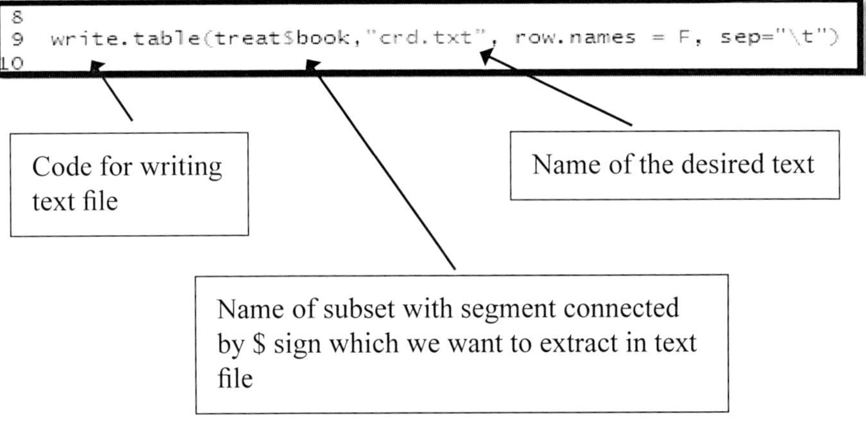

To view the text file, either locate it in your working directory or use the following code to open it directly:

file.show("crd.txt")

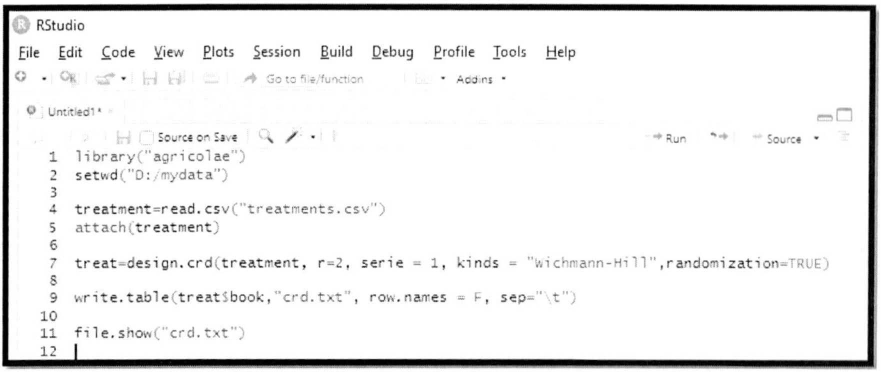

RStudio

File Edit Code View Plots Session Build Debug Profile Tools Help

Untitled1*

```
library("agricolae")
setwd("D:/mydata")

treatment=read.csv("treatments.csv")
attach(treatment)

treat=design.crd(treatment, r=2, serie = 1, kinds = "Wichmann-Hill",randomization=TRUE)

write.table(treat$book,"crd.txt", row.names = F, sep="\t")

file.show("crd.txt")
```

Output in crd.txt file

| "plots" | "r" | "treatment" |
|---|---|---|
| 11 | 1 | "T20" |
| 12 | 1 | "T6" |
| 13 | 1 | "T7" |
| 14 | 1 | "T3" |
| 15 | 1 | "T10" |
| 16 | 2 | "T10" |
| 17 | 1 | "T2" |
| 18 | 1 | "T9" |
| 19 | 1 | "T14" |
| 20 | 1 | "T16" |
| 21 | 1 | "T5" |
| 22 | 1 | "T18" |
| 23 | 1 | "T17" |
| 24 | 2 | "T6" |
| 25 | 1 | "T1" |
| 26 | 2 | "T17" |

| | | |
|---|---|---|
| 27 | 2 | "T16" |
| 28 | 1 | "T19" |
| 29 | 2 | "T5" |
| 30 | 2 | "T19" |
| 31 | 2 | "T3" |
| 32 | 1 | "T12" |
| 33 | 2 | "T1" |
| 34 | 1 | "T4" |
| 35 | 2 | "T20" |
| 36 | 1 | "T13" |
| 37 | 1 | "T8" |
| 38 | 2 | "T4" |
| 39 | 2 | "T18" |
| 40 | 1 | "T15" |
| 41 | 2 | "T12" |
| 42 | 2 | "T2" |
| 43 | 2 | "T9" |
| 44 | 2 | "T13" |
| 45 | 2 | "T8" |
| 46 | 1 | "T11" |
| 47 | 2 | "T7" |
| 48 | 2 | "T14" |
| 49 | 2 | "T15" |
| 50 | 2 | "T11" |

Here's how the output appears in the text file

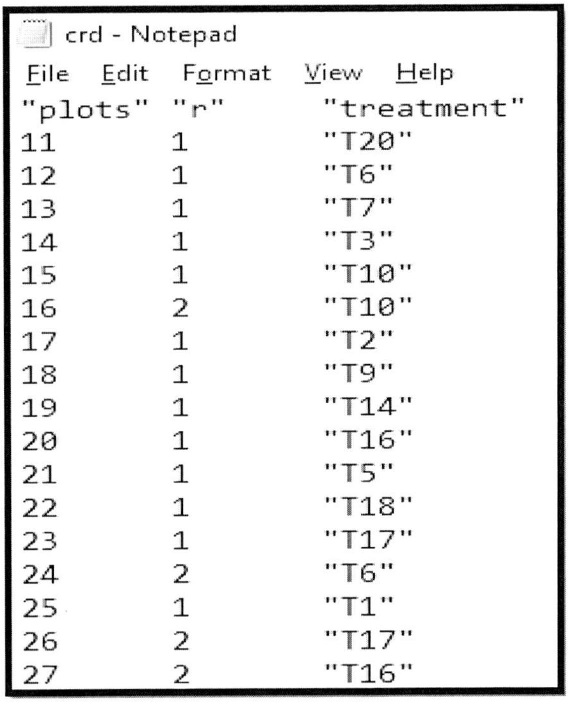

crd - Notepad

File Edit Format View Help

```
"plots" "r"      "treatment"
11        1      "T20"
12        1      "T6"
13        1      "T7"
14        1      "T3"
15        1      "T10"
16        2      "T10"
17        1      "T2"
18        1      "T9"
19        1      "T14"
20        1      "T16"
21        1      "T5"
22        1      "T18"
23        1      "T17"
24        2      "T6"
25        1      "T1"
26        2      "T17"
27        2      "T16"
```

This output represents the randomized treatment assignments for a CRD experiment. It specifies which treatments are assigned to each experimental unit.

For example:

first row (Row 1) shows that unit number 11 belongs to repetition 1 ($r = 1$) and has been assigned treatment T16.

## Randomization for RBD

i. Begin with an input file in the specified format

ii. Read this file in RStudio using the command:

trt = read.csv(“treatments.csv”)

✓ Note: The treatment vector in the file contains the various names of treatments that will be applied in the field.

iii. To randomize these treatments, follow these steps:

- Attach your file using the attach() command.
- Extract only the treatment names using the following code:

trt_RCBD = trt[1:20,]

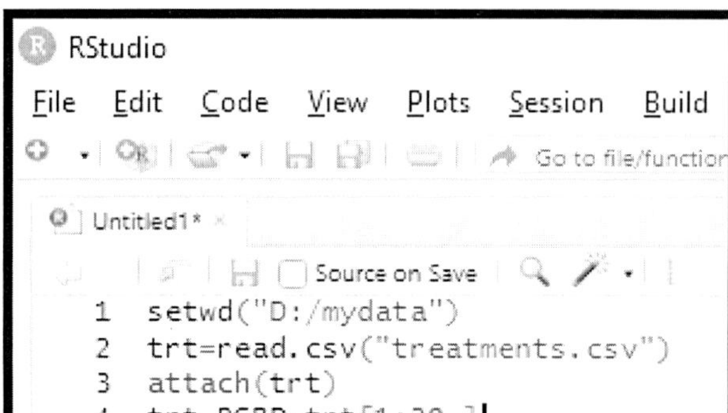


[1:20,] It read the rows from the “trt” subset and start from the second row to avoid title.

This subset is saved into trt_RCBD

**In the console window, it will look like this**

```
Console  Terminal  Jobs
R 4.0.2 · D:/mydata/
> trt_RCBD=trt[1:20,]
> trt_RCBD
 [1] "T1"  "T2"  "T3"  "T4"  "T5"  "T6"  "T7"  "T8"  "T9"  "T10" "T11" "T12" "T13" "T14" "T15"
[16] "T16" "T17" "T18" "T19" "T20"
>
```

Write the following code in the script editor:

outdesign = design.rcbd (trt, r=3,serie=1986, “Wichmann-Hill,” first = F)

```
setwd("D:/mydata")
trt=read.csv("treatments.csv")
attach(trt)
trt_RCBD=trt[1:20,]

library("agricolae")

outdesign = design.rcbd(trt_RCBD, r=3,serie=1,986,"Wichmann-Hill", first = F)
outdesign
```

This code will generate an “outdesign” variable containing the randomized complete block design (RCBD) based on your treatment data stored in the “trt” variable.

**Description of code**

- **design.rcbd**: Randomizes treatments for RCBD.
- **trt_RCBD**: Subset of treatment names
- **r**: Replications or blocks
- **serie**: Plot numbering: 1: 11,12; 2: 101,102; 3: 1001,1002
- **seed**: Refers to the seed value for randomization
- **kinds**: Randomization method (“Wichmann-Hill,” “Marsaglia-Multicarry,” “Super-Duper,” “Mersenne-Twister,” “Knuth-TAOCP,” “user-supplied,” “Knuth-TAOCP-2002,” “default”)
- **first**: Randomize rep 1 (TRUE/FALSE)

**Output in console**

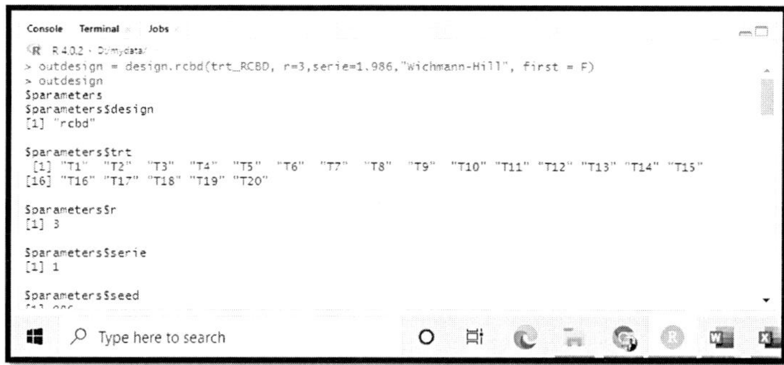

```
> outdesign = design.rcbd(trt_RCBD, r=3,serie=1,986,"Wichmann-Hill", first = F)
> outdesign
$parameters
$parameters$design
[1] "rcbd"

$parameters$trt
 [1] "T1"  "T2"  "T3"  "T4"  "T5"  "T6"  "T7"  "T8"  "T9"  "T10" "T11" "T12" "T13" "T14" "T15"
[16] "T16" "T17" "T18" "T19" "T20"

$parameters$r
[1] 3

$parameters$serie
[1] 1

$parameters$seed
```

The console displays detailed information about the parameters used in the code. To save this entire console output to a text file, you can use the "sink" code.

In this output we can see the addition in the environment

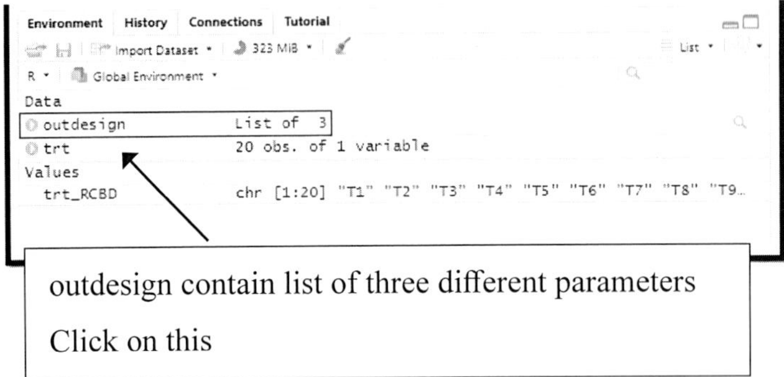


After clicking on this bar, a new script will open in the script editor. This script describes the "outdesign" as follows:

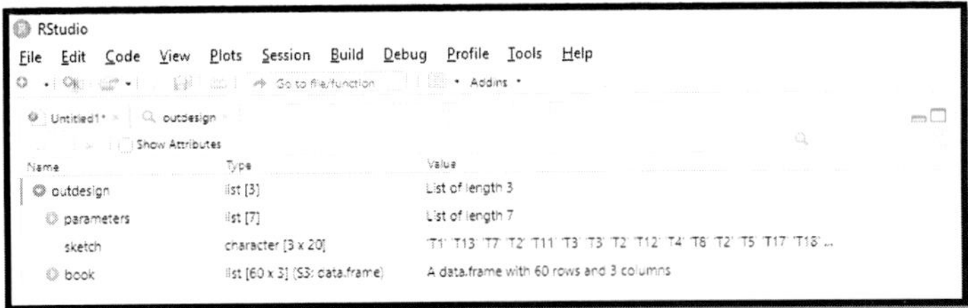


Subset "outdesign" contain "parameters," "sketch" and "book"

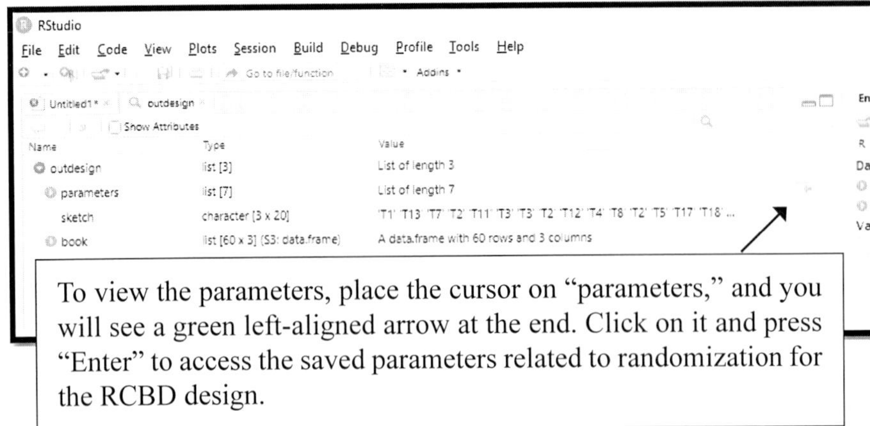

If you wish to create a sketch of the randomization in a Word (.doc format) file, you need to download and launch a package called "bruceR." Then, follow the steps outlined below:

```
setwd("D:/mydata")
trt=read.csv("treatments.csv")
attach(trt)
trt_RCBD=trt[1:20,]

library("agricolae")

outdesign = design.rcbd(trt_RCBD, r=3,serie=1,986,"Wichmann-Hill", first = F)
outdesign

library("bruceR")
print_table(outdesign$sketch, file="Field layout")
```

Write the code provided in the script editor to generate the sketch.

```
library("bruceR")
print_table(outdesign$sketch, file="Field layout")
```

**Code description**

- **print_table** : Writes a table in a Word file.
- **outdesign$sketch** : Retrieves field layout
- **file= "Field layout"** : Names the Word file as "Field layout"

To see the result, run this code and check the console for the output

```
R 4.0.2 · D:/mydata/
✓ data.table

Key functions of `bruceR`:
Basic           Variable        Modeling
set_wd()        Describe()      TTEST()
import()        Freq()          MANOVA()
export()        Corr()          EMMEANS()
print_table()   Alpha()         PROCESS()
MEAN()          EFA()           model_summary()
LOOKUP()        CFA()           lavaan_summary()

Warning message:
package 'bruceR' was built under R version 4.0.5
> print_table(outdesign$sketch, file="Field layout")
✓ Table saved to "D:/mydata/Field layout.doc"

>
```

The table saved to "D:/mydata/Field layout.doc" and verify it.

Similarly, we can save the "book" from the "outdesign" for collecting field data directly. Here's how to create the "book" using the following code:

```
library("bruceR")
print_table(outdesign$book, file="Field book")
```

**Output in the word file**

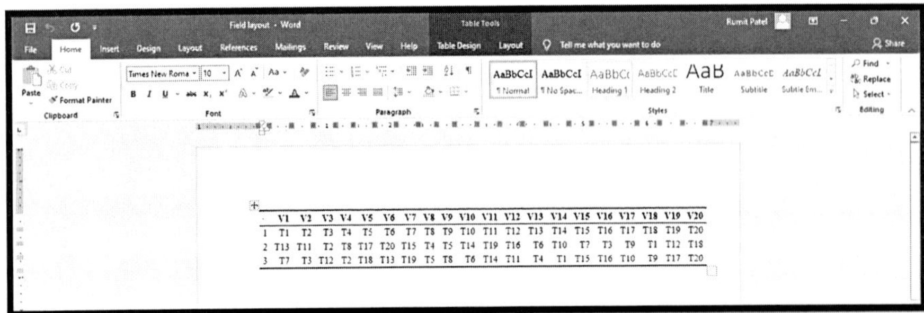

|  | V1 | V2 | V3 | V4 | V5 | V6 | V7 | V8 | V9 | V10 | V11 | V12 | V13 | V14 | V15 | V16 | V17 | V18 | V19 | V20 |
|---|---|---|---|---|---|---|---|---|---|---|---|---|---|---|---|---|---|---|---|---|
| 1 | T1 | T2 | T3 | T4 | T5 | T6 | T7 | T8 | T9 | T10 | T11 | T12 | T13 | T14 | T15 | T16 | T17 | T18 | T19 | T20 |
| 2 | T13 | T11 | T2 | T8 | T17 | T20 | T15 | T4 | T5 | T14 | T19 | T16 | T6 | T10 | T7 | T3 | T9 | T1 | T12 | T18 |
| 3 | T7 | T3 | T12 | T2 | T18 | T13 | T19 | T5 | T8 | T6 | T14 | T11 | T4 | T1 | T15 | T16 | T10 | T9 | T17 | T20 |

|  | plots | block | trt_RCBD |
|---|---|---|---|
| 1 | 11.000 | 1 | T1 |
| 2 | 12.000 | 1 | T2 |
| 3 | 13.000 | 1 | T3 |
| 4 | 14.000 | 1 | T4 |
| 5 | 15.000 | 1 | T5 |
| 6 | 16.000 | 1 | T6 |
| 7 | 17.000 | 1 | T7 |
| 8 | 18.000 | 1 | T8 |
| 9 | 19.000 | 1 | T9 |
| 10 | 20.000 | 1 | T10 |
| 11 | 21.000 | 1 | T11 |
| 12 | 22.000 | 1 | T12 |
| 13 | 23.000 | 1 | T13 |
| 14 | 24.000 | 1 | T14 |
| 15 | 25.000 | 1 | T15 |
| 16 | 26.000 | 1 | T16 |
| 17 | 27.000 | 1 | T17 |

| | plots | block | trt_RCBD |
|---|---|---|---|
| 18 | 28.000 | 1 | T18 |
| 19 | 29.000 | 1 | T19 |
| 20 | 30.000 | 1 | T20 |
| 21 | 21.000 | 2 | T13 |
| 22 | 22.000 | 2 | T11 |
| 23 | 23.000 | 2 | T2 |
| 24 | 24.000 | 2 | T8 |
| 25 | 25.000 | 2 | T17 |
| 26 | 26.000 | 2 | T20 |
| 27 | 27.000 | 2 | T15 |
| 28 | 28.000 | 2 | T4 |
| 29 | 29.000 | 2 | T5 |
| 30 | 30.000 | 2 | T14 |
| 31 | 31.000 | 2 | T19 |
| 32 | 32.000 | 2 | T16 |
| 33 | 33.000 | 2 | T6 |
| 34 | 34.000 | 2 | T10 |
| 35 | 35.000 | 2 | T7 |
| 36 | 36.000 | 2 | T3 |
| 37 | 37.000 | 2 | T9 |
| 38 | 38.000 | 2 | T1 |
| 39 | 39.000 | 2 | T12 |
| 40 | 40.000 | 2 | T18 |
| 41 | 31.000 | 3 | T7 |
| 42 | 32.000 | 3 | T3 |
| 43 | 33.000 | 3 | T12 |
| 44 | 34.000 | 3 | T2 |
| 45 | 35.000 | 3 | T18 |
| 46 | 36.000 | 3 | T13 |
| 47 | 37.000 | 3 | T19 |
| 48 | 38.000 | 3 | T5 |
| 49 | 39.000 | 3 | T8 |
| 50 | 40.000 | 3 | T6 |
| 51 | 41.000 | 3 | T14 |
| 52 | 42.000 | 3 | T11 |

| | plots | block | trt_RCBD |
|---|---|---|---|
| 53 | 43.000 | 3 | T4 |
| 54 | 44.000 | 3 | T1 |
| 55 | 45.000 | 3 | T15 |
| 56 | 46.000 | 3 | T16 |
| 57 | 47.000 | 3 | T10 |
| 58 | 48.000 | 3 | T9 |
| 59 | 49.000 | 3 | T17 |
| 60 | 50.000 | 3 | T20 |

## Randomization for LSD

Randomization for Latin square design (LSD) involves arranging t-treatments, each repeated t-times so that every treatment appears once in each row and column. Follow these steps for randomization:

i. Prepare an input file in .csv format with the names of treatments.

**Input file in .csv format**

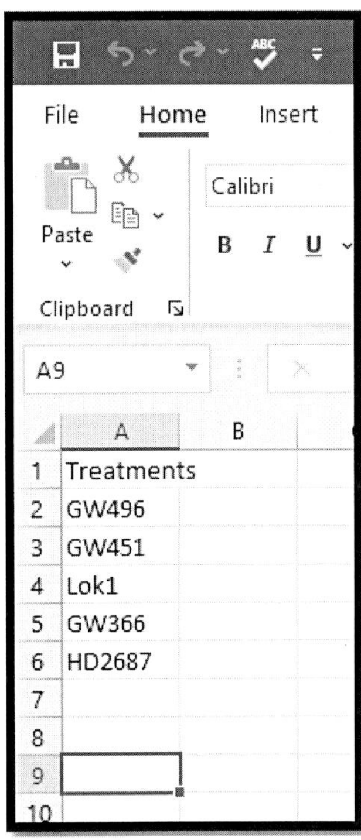


Here, this is an input file with the name of treatments which is to be evaluated in field using Latin square design

ii. Set your working directory.

iii. Read the input file in RStudio as explained earlier.

iv. Follow the instructions below for randomization.

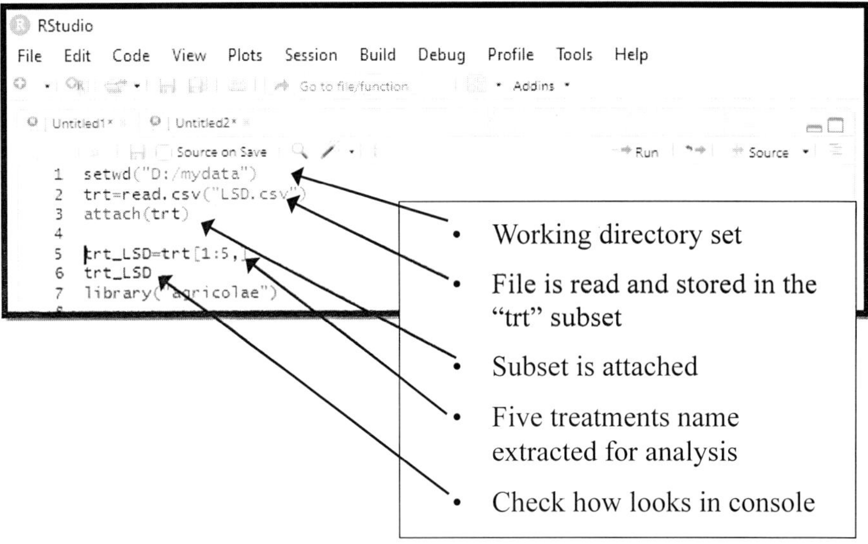


**Console**

```
Console   Terminal   Jobs

R 4.0.2 · D:/mydata/
> setwd("D:/mydata")
> trt=read.csv("LSD.csv")
> attach(trt)
The following object is masked from trt (pos = 3):

    Treatments

> trt_LSD=trt[1:5,]
> trt_LSD
[1] "GW496"  "GW451"  "Lok1"   "GW366"  "HD2687"
>
```

v. Launch the "agricolae" package.

vi. Apply the following code to randomize all five treatments in LSD

outdesign = design.lsd(trt_LSD, serie=2, seed=23)

**In script editor**

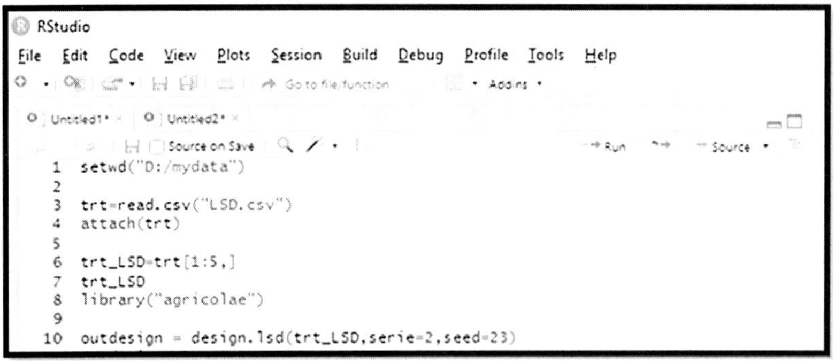

```
setwd("D:/mydata")

trt=read.csv("LSD.csv")
attach(trt)

trt_LSD=trt[1:5,]
trt_LSD
library("agricolae")

outdesign = design.lsd(trt_LSD,serie=2,seed=23)
```

vii. Press Ctrl+Enter (Run) and see the output in the console

We find the output in different sets like $parameters, $sketch, $book

To save the sketch and book outputs in a Word file, use the following code:

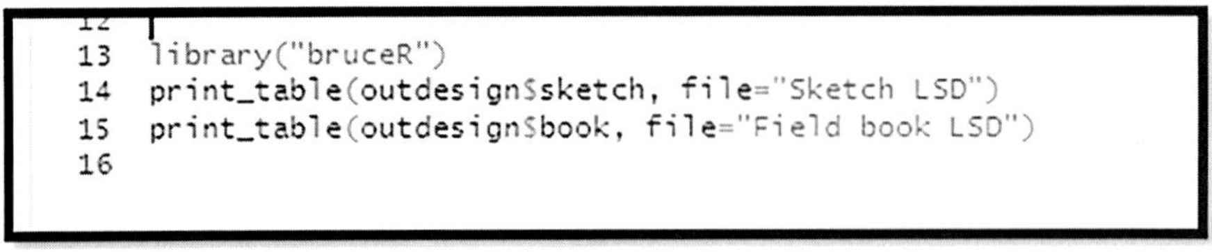

```
library("bruceR")
print_table(outdesign$sketch, file="Sketch LSD")
print_table(outdesign$book, file="Field book LSD")
```

Package "bruceR" help us to draw the sketch and field book in the working directory in .doc format

Console looks like

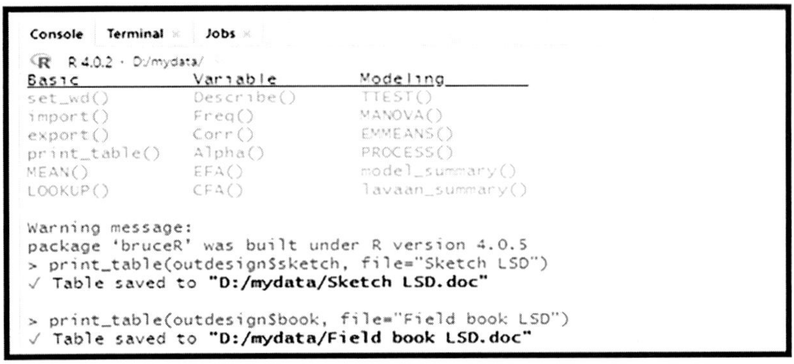

```
R 4.0.2 · D:/mydata/
Basic          Variable       Modeling
set_wd()       Describe()     TTEST()
import()       Freq()         MANOVA()
export()       Corr()         EMMEANS()
print_table()  Alpha()        PROCESS()
MEAN()         EFA()          model_summary()
LOOKUP()       CFA()          lavaan_summary()

Warning message:
package 'bruceR' was built under R version 4.0.5
> print_table(outdesign$sketch, file="Sketch LSD")
✓ Table saved to "D:/mydata/Sketch LSD.doc"

> print_table(outdesign$book, file="Field book LSD")
✓ Table saved to "D:/mydata/Field book LSD.doc"
```

Output for sketch and field book in word file

**#sketch**

| | V1 | V2 | V3 | V4 | V5 |
|---|---|---|---|---|---|
| 1 | GW496 | GW366 | GW451 | Lok1 | HD2687 |
| 2 | GW366 | GW451 | HD2687 | GW496 | Lok1 |
| 3 | HD2687 | Lok1 | GW496 | GW451 | GW366 |
| 4 | GW451 | HD2687 | Lok1 | GW366 | GW496 |
| 5 | Lok1 | GW496 | GW366 | HD2687 | GW451 |

**# Field book**

| | plots | row | col | trt_LSD |
|---|---|---|---|---|
| 1 | 101.000 | 1 | 1 | GW496 |
| 2 | 102.000 | 1 | 2 | GW366 |
| 3 | 103.000 | 1 | 3 | GW451 |
| 4 | 104.000 | 1 | 4 | Lok1 |
| 5 | 105.000 | 1 | 5 | HD2687 |
| 6 | 201.000 | 2 | 1 | GW366 |
| 7 | 202.000 | 2 | 2 | GW451 |
| 8 | 203.000 | 2 | 3 | HD2687 |
| 9 | 204.000 | 2 | 4 | GW496 |
| 10 | 205.000 | 2 | 5 | Lok1 |
| 11 | 301.000 | 3 | 1 | HD2687 |
| 12 | 302.000 | 3 | 2 | Lok1 |
| 13 | 303.000 | 3 | 3 | GW496 |
| 14 | 304.000 | 3 | 4 | GW451 |
| 15 | 305.000 | 3 | 5 | GW366 |
| 16 | 401.000 | 4 | 1 | GW451 |
| 17 | 402.000 | 4 | 2 | HD2687 |
| 18 | 403.000 | 4 | 3 | Lok1 |
| 19 | 404.000 | 4 | 4 | GW366 |
| 20 | 405.000 | 4 | 5 | GW496 |
| 21 | 501.000 | 5 | 1 | Lok1 |
| 22 | 502.000 | 5 | 2 | GW496 |
| 23 | 503.000 | 5 | 3 | GW366 |
| 24 | 504.000 | 5 | 4 | HD2687 |
| 25 | 505.000 | 5 | 5 | GW451 |

## Randomization for Split Plot Design (SPD)

For the split-plot design, two stages of randomization are employed to allocate experimental units to treatments. In the first stage, experimental units are randomized to levels of treatment factor A across the entire plot. Then, the plot is subdivided into subplots. In the second stage, experimental units within each subplot are randomized to levels of treatment factor B.

### Input file

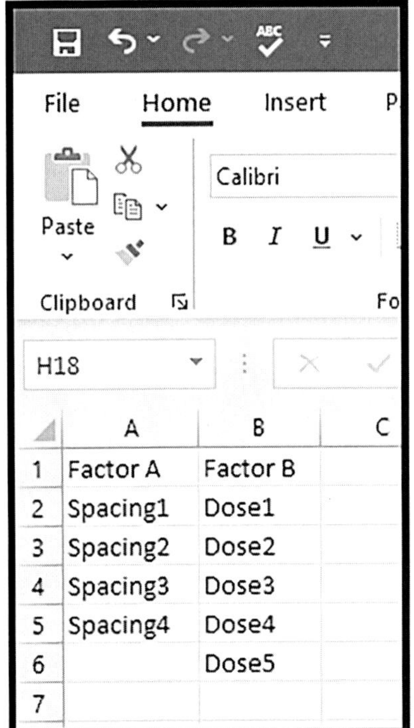

| | A | B | C |
|---|---|---|---|
| 1 | Factor A | Factor B | |
| 2 | Spacing1 | Dose1 | |
| 3 | Spacing2 | Dose2 | |
| 4 | Spacing3 | Dose3 | |
| 5 | Spacing4 | Dose4 | |
| 6 | | Dose5 | |
| 7 | | | |

Here, the input file indicates the factors "Factor A" and "Factor B"

Factor A contain four treatments and factor B contain five treatments.

Now save this file in .csv format

Read this file in RStudio as described in an earlier chapter below:

```
setwd("D:/mydata")
input=read.csv("trt_SPD.csv")
input
```

Now check the input subset in the console

**Console**

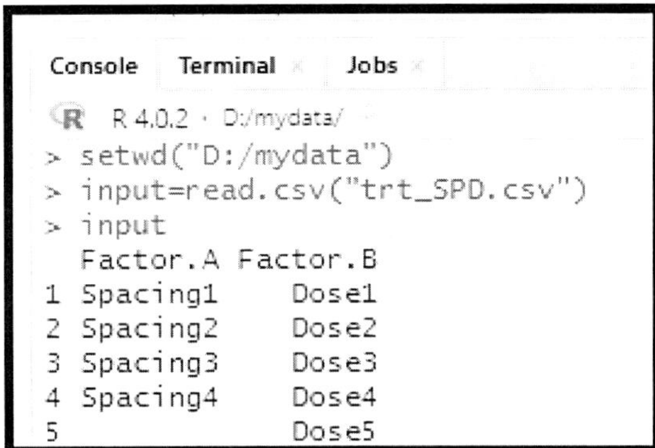


To extract the Factor A treatments and create a subset, use the following code in the script editor:

```
trt_a=input$Factor.A[1:4]
```

write the above code on the script editor and check the subset "trt_a" in the console

**In Script editor**

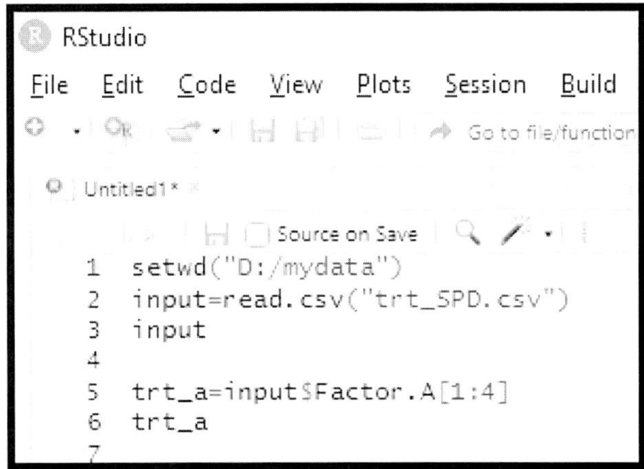

## Output in console

```
Console    Terminal    Jobs
R 4.0.2 · D:/mydata/
> setwd("D:/mydata")
> input=read.csv("trt_SPD.csv")
> input
  Factor.A Factor.B
1 Spacing1    Dose1
2 Spacing2    Dose2
3 Spacing3    Dose3
4 Spacing4    Dose4
5             Dose5
> trt_a=input$Factor.A[1:4]
> trt_a
[1] "Spacing1" "Spacing2" "Spacing3" "Spacing4"
>
```

Similarly, do for Factor B and save it in "trt_b" and run the randomization code for split-plot design like below:

outdesign=design.split(trt_a, trt_b,r=3,serie=2,seed=45,kinds = "Super-Duper")

```
RStudio
File Edit Code View Plots Session Build Debug Profile Tools Help
Untitled1*
setwd("D:/mydata")
input=read.csv("trt_SPD.csv")
input

trt_a=input$Factor.A[1:4]
trt_a
trt_b=input$Factor.B[1:5]
trt_b

outdesign=design.split(trt_a, trt_b,r=3,serie=2,seed=45,kinds ="Super-Duper"
outdesign
```

Now save this output in a Word file using the package "bruceR" as described in randomization for RCBD.

Here, this code gives only the "book" not a "layout" or "sketch" of the field

**Check the output into the word file**

| | Plots | Splots | Block | trt_a | trt_b |
|---|---|---|---|---|---|
| 1 | 101.000 | 1 | 1 | Spacing3 | Dose1 |
| 2 | 101.000 | 2 | 1 | Spacing3 | Dose4 |
| 3 | 101.000 | 3 | 1 | Spacing3 | Dose5 |
| 4 | 101.000 | 4 | 1 | Spacing3 | Dose2 |
| 5 | 101.000 | 5 | 1 | Spacing3 | Dose3 |
| 6 | 102.000 | 1 | 1 | Spacing1 | Dose1 |
| 7 | 102.000 | 2 | 1 | Spacing1 | Dose4 |
| 8 | 102.000 | 3 | 1 | Spacing1 | Dose2 |
| 9 | 102.000 | 4 | 1 | Spacing1 | Dose3 |
| 10 | 102.000 | 5 | 1 | Spacing1 | Dose5 |
| 11 | 103.000 | 1 | 1 | Spacing4 | Dose1 |
| 12 | 103.000 | 2 | 1 | Spacing4 | Dose5 |
| 13 | 103.000 | 3 | 1 | Spacing4 | Dose4 |
| 14 | 103.000 | 4 | 1 | Spacing4 | Dose2 |
| 15 | 103.000 | 5 | 1 | Spacing4 | Dose3 |
| 16 | 104.000 | 1 | 1 | Spacing2 | Dose4 |
| 17 | 104.000 | 2 | 1 | Spacing2 | Dose3 |
| 18 | 104.000 | 3 | 1 | Spacing2 | Dose1 |
| 19 | 104.000 | 4 | 1 | Spacing2 | Dose5 |
| 20 | 104.000 | 5 | 1 | Spacing2 | Dose2 |
| 21 | 105.000 | 1 | 2 | Spacing3 | Dose1 |
| 22 | 105.000 | 2 | 2 | Spacing3 | Dose3 |
| 23 | 105.000 | 3 | 2 | Spacing3 | Dose5 |
| 24 | 105.000 | 4 | 2 | Spacing3 | Dose4 |
| 25 | 105.000 | 5 | 2 | Spacing3 | Dose2 |
| 26 | 106.000 | 1 | 2 | Spacing2 | Dose1 |
| 27 | 106.000 | 2 | 2 | Spacing2 | Dose2 |
| 28 | 106.000 | 3 | 2 | Spacing2 | Dose5 |
| 29 | 106.000 | 4 | 2 | Spacing2 | Dose4 |
| 30 | 106.000 | 5 | 2 | Spacing2 | Dose3 |
| 31 | 107.000 | 1 | 2 | Spacing4 | Dose1 |
| 32 | 107.000 | 2 | 2 | Spacing4 | Dose3 |
| 33 | 107.000 | 3 | 2 | Spacing4 | Dose2 |
| 34 | 107.000 | 4 | 2 | Spacing4 | Dose4 |

| | Plots | Splots | Block | trt_a | trt_b |
|---|---|---|---|---|---|
| 35 | 107.000 | 5 | 2 | Spacing4 | Dose5 |
| 36 | 108.000 | 1 | 2 | Spacing1 | Dose3 |
| 37 | 108.000 | 2 | 2 | Spacing1 | Dose5 |
| 38 | 108.000 | 3 | 2 | Spacing1 | Dose1 |
| 39 | 108.000 | 4 | 2 | Spacing1 | Dose2 |
| 40 | 108.000 | 5 | 2 | Spacing1 | Dose4 |
| 41 | 109.000 | 1 | 3 | Spacing4 | Dose1 |
| 42 | 109.000 | 2 | 3 | Spacing4 | Dose3 |
| 43 | 109.000 | 3 | 3 | Spacing4 | Dose2 |
| 44 | 109.000 | 4 | 3 | Spacing4 | Dose5 |
| 45 | 109.000 | 5 | 3 | Spacing4 | Dose4 |
| 46 | 110.000 | 1 | 3 | Spacing3 | Dose1 |
| 47 | 110.000 | 2 | 3 | Spacing3 | Dose4 |
| 48 | 110.000 | 3 | 3 | Spacing3 | Dose5 |
| 49 | 110.000 | 4 | 3 | Spacing3 | Dose3 |
| 50 | 110.000 | 5 | 3 | Spacing3 | Dose2 |
| 51 | 111.000 | 1 | 3 | Spacing1 | Dose4 |
| 52 | 111.000 | 2 | 3 | Spacing1 | Dose3 |
| 53 | 111.000 | 3 | 3 | Spacing1 | Dose2 |
| 54 | 111.000 | 4 | 3 | Spacing1 | Dose5 |
| 55 | 111.000 | 5 | 3 | Spacing1 | Dose1 |
| 56 | 112.000 | 1 | 3 | Spacing2 | Dose5 |
| 57 | 112.000 | 2 | 3 | Spacing2 | Dose3 |
| 58 | 112.000 | 3 | 3 | Spacing2 | Dose4 |
| 59 | 112.000 | 4 | 3 | Spacing2 | Dose1 |
| 60 | 112.000 | 5 | 3 | Spacing2 | Dose2 |

✓ Note:

Same way, strip plot design is also randomized with the same subsets of treatments using the code below

```
design.strip(trt_a, trt_b ,r=3, serie=2,seed=45,kinds = "Super-Duper")
```

The output will remain the same like split-plot design

# 5

# Correlation

This chapter begins with an introductory overview, providing a basic understanding of correlation and its types. It guides readers through in-depth analysis using R, demonstrating how to apply these concepts practically.

## Correlation

Correlation analysis is a statistical method used to evaluate the strength and direction of the relationship between two variables. This analysis is vital in identifying and understanding how variables are interconnected within a dataset, providing insights into their mutual influences.

Key aspects of correlation include:

- Range of the correlation coefficient

  The value of the correlation coefficient ("*r*") falls within the range of −1 to +1.

- Independence from origin and scale

  The correlation coefficient ("*r*") remains consistent regardless of any changes in the origin or scale of the variables.

- Unit independence

  The correlation coefficient is independent of the units of measurement of the variables.

- Relationship with independence

  While uncorrelated variables are generally independent, the reverse is not necessarily true; independent variables will always be uncorrelated.

## Types of Correlation

- Positive correlation

  When two variables increase or decrease together. For instance, between yield and number of secondary branches in moong.

- Negative correlation

  When one variable increases while the other decreases. In wheat breeding, for example, negative correlation between plant height with yield.

- Simple correlation

  It is the relationship between two variables without the influence of other variables. An example is the correlation between seed size and seedling vigor in maize; larger seeds may lead to more vigorous seedlings.

- Partial correlation

  This type assesses the relationship between two variables while controlling for the effect of one or more other variables. For instance, in rice breeding, the correlation between plant height and yield might be analyzed while controlling for water availability.

- Multiple correlation

  This involves understanding the relationship between one variable and a combination of two or more other variables. In grape breeding, this could be the relationship between vine yield and a combination of factors like sunlight exposure, soil pH, and pruning practices.

## Correlation Coefficients and Its Test

### *Techniques for Measuring Correlation*

Three important statistical tools used to measure correlation are Pearson's correlation, Kendall's correlation, and Spearman's rank correlation.

Karl Pearson suggested a mathematical method for measuring the magnitude of the linear relationship between the two variables. It is the most widely used method in practice and is denoted by "*r*." The formula for calculating "*r*" is when $x$ and $y$ are two characters:

$$r_{xy} = \frac{Cov(x, y)}{\sqrt{V(x)\,V(y)}},$$

### *Test for Significance of Correlation Coefficient*

If "*r*" is the observed correlation coefficient in a sample of "*n*" pairs of observations from a bivariate normal population, then Prof. Fisher proved that under the null hypothesis,

$$H_0: \rho = 0,$$

$$H_a: \rho \neq 0.$$

If the population correlation coefficient of $x$ and $y$ is denoted by $\rho$, then it is often of interest to test whether $\rho$ is zero or different from zero, based on the observed correlation coefficient "$r$."

Thus if "$r$" is the sample correlation coefficient based on a sample of "$n$" observations, then the appropriate test statistic for testing the null hypothesis against the alternative hypothesis is:

$$t = \frac{|r|}{\sqrt{1-r^2}}\sqrt{n-2} \text{ follows Student's } t\text{-distribution with } (n-2) \text{ d.f.}$$

If the calculated value of $t >$ table value of $t$ with $(n - 2)$ d.f. at a specified level of significance, then the null hypothesis is rejected. That is, there may be a significant correlation between the two variables. Otherwise, the null hypothesis is accepted.

## Correlation Analysis in R

In RStudio, it's beneficial to set up a working directory for easier data handling. We're analyzing a non-replicated tomato population dataset, suitable for certain correlation methods. For plant breeding research, however, replicated data is typically used to study genotypic and phenotypic correlations.

### Data format in .csv file

| Plant No | PH | NPB | CHL | Phenol | Moisture | DS |
|---|---|---|---|---|---|---|
| 1 | 254 | 4 | 7.2251 | 2.489628 | 76.7285 | 0 |
| 2 | 265 | 3 | 7.59076 | 1.249025 | 84.42953 | 1 |
| 3 | 180 | 5 | 8.43912 | 2.067269 | 80.27613 | 0 |
| 4 | 298 | 3 | 8.71716 | 2.51249 | 79.97725 | 0 |
| 5 | 265 | 5 | 7.67898 | 1.885571 | 80.77361 | 0 |
| 6 | 265 | 3 | 9.12202 | 1.417487 | 80.26667 | 0 |
| 7 | 245 | 5 | 14.55768 | 1.713499 | 76.65317 | 0 |
| 8 | 270 | 3 | 11.91872 | 1.961379 | 78.36364 | 0 |
| 9 | 200 | 4 | 17.33508 | 1.415081 | 72.99185 | 0 |
| 10 | 296 | 6 | 11.89406 | 2.731491 | 73.82813 | 0 |
| 11 | 290 | 5 | 12.21606 | 0.86036 | 71.06825 | 2 |
| 12 | 278 | 3 | 6.7204 | 0.890442 | 83.78906 | 2 |
| 13 | 190 | 7 | 10.63072 | 1.611219 | 77.55396 | 0 |
| 14 | 245 | 4 | 7.18706 | 1.428 | 76.29 | 0 |
| 15 | 240 | 3 | 14.49648 | 2.105775 | 74.0625 | 0 |
| 16 | 265 | 6 | 10.43764 | 1.937313 | 77.50611 | 0 |

To examine the structure of a dataset in R, apply the "str()" function, like str(name_of_data). This command provides a detailed overview of the dataset's composition.

```
RStudio
File Edit Code View Plots Session Build Debug Profile Tools Help
Go to file/function    Addins
Untitled1*
Source on Save
setwd("D:/mydata")
data=read.csv("tomato.csv")

str(data)
```

**Console window**

```
Console  Terminal  Jobs
R 4.0.2 · D:/mydata/
> setwd("D:/mydata")
> data=read.csv("tomato.csv")
> str(data)
'data.frame':	235 obs. of  7 variables:
 $ Plant.No: int  1 2 3 4 5 6 7 8 9 10 ...
 $ PH      : int  254 265 180 298 265 265 245 270 200 296 ...
 $ NPB     : int  4 3 5 3 5 3 5 3 4 6 ...
 $ CHL     : num  7.23 7.59 8.44 8.72 7.68 ...
 $ Phenol  : num  2.49 1.25 2.07 2.51 1.89 ...
 $ Moisture: num  76.7 84.4 80.3 80 80.8 ...
 $ DS      : int  0 1 0 0 0 0 0 0 0 0 ...
>
```

In our dataset, "Plant No" is not a variable for correlation analysis. Instead, we focus on six variables located in columns two to seven of the .csv file. To access these, create a data frame using "data[2:7]", where "data" is the subset name and [2:7] specifies columns 2–7 containing our relevant variables.

### *Measurement of Correlation*

To perform a correlation analysis in RStudio, use the following code:

correlation = cor(data[2:7], method = "pearson")

- Here, "correlation" is a new variable where the results are stored
- The "cor" function, included in R's basic package, conducts the analysis

- data[2:7] specifies the dataset's columns 2 to 7, containing the variables of interest
- The method="pearson" argument indicates the use of Pearson's correlation method
- To use alternative methods, such as Kendall or Spearman's rank correlation, simply replace "pearson" with "kendall," or "Spearman" in the code

```
RStudio
File  Edit  Code  View  Plots  Session  Build  Debug  Profile
Go to file/function  Addins
Untitled1*
Source on Save
setwd("D:/mydata")
data=read.csv("tomato.csv")

correlation=cor(data[2:7], method = "pearson")
```

To view the results of the correlation analysis in RStudio, first run the code. This saves the results in the subset named "correlation." Then, to see the output, simply type "correlation" in the console and press "Enter." The results will be displayed in the console window.

**Console**

```
Console  Terminal  Jobs
R 4.0.2 · D:/mydata/
> setwd("D:/mydata")
> data=read.csv("tomato.csv")
> correlation=cor(data[2:7], method = "pearson")
> correlation
                   PH         NPB         CHL       Phenol     Moisture           DS
PH        1.000000000  0.54387761 -0.03614283 -0.030777598 -0.094881217  0.002121988
NPB       0.543877614  1.00000000 -0.05336287 -0.012145602 -0.040113832 -0.027004502
CHL      -0.036142831 -0.05336287  1.00000000  0.122403828 -0.032099362 -0.090900577
Phenol   -0.030777598 -0.01214560  0.12240383  1.000000000 -0.002576872 -0.818698226
Moisture -0.094881217 -0.04011383 -0.03209936 -0.002576872  1.000000000 -0.059000293
DS        0.002121988 -0.02700450 -0.09090058 -0.818698226 -0.059000293  1.000000000
> |
```

To save the output of the correlation analysis in a Word document (.doc) in a table format, use the "bruceR" package as outlined in the relevant chapter. This package allows for efficient exporting of results directly into a .doc file, presenting the data in a well-organized table layout.

| | PH | NPB | CHL | Phenol | Moisture | DS |
|---|---|---|---|---|---|---|
| PH | 1.000 | 0.544 | –0.036 | –0.031 | –0.095 | 0.002 |
| NPB | 0.544 | 1.000 | –0.053 | –0.012 | 0.040 | –0.027 |
| CHL | –0.036 | –0.053 | 1.000 | 0.122 | –0.032 | –0.091 |
| Phenol | –0.031 | –0.012 | 0.122 | 1.000 | –0.003 | –0.819 |
| Moisture | –0.095 | –0.040 | –0.032 | –0.003 | 1.000 | –0.059 |
| DS | 0.002 | –0.027 | –0.091 | –0.819 | –0.059 | 1.000 |

To test the significance of correlation values in R, a *t*-test is required, which involves using the "Hmisc" package. Install this package and then use it for conducting the significance test of the correlation. The code for this analysis is as follows:

```
test=rcorr(as.matrix(data[2:7]))
```

This code utilizes the rcorr function from the "Hmisc" package. It computes a matrix of either Pearson's *r* or Spearman's rho rank correlation coefficients for every possible pair of columns in a matrix. The function is designed to handle missing values efficiently by deleting them in pairs, which avoids the need to remove entire rows that have any missing elements.

### Script editor

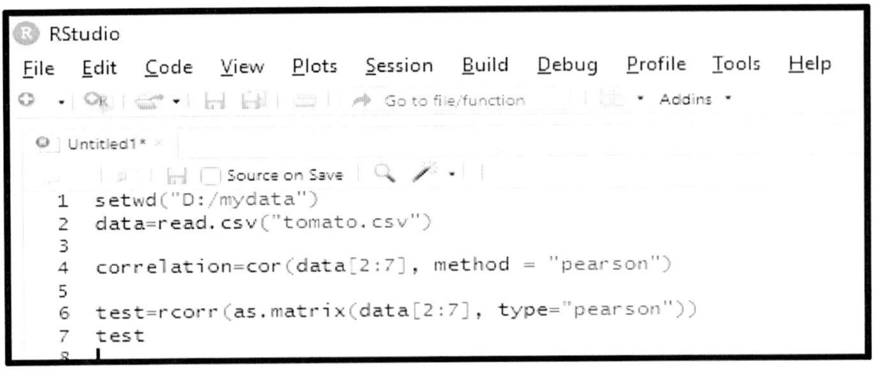

RStudio

File Edit Code View Plots Session Build Debug Profile Tools Help

Go to file/function Addins

Untitled1*

Source on Save

```
setwd("D:/mydata")
data=read.csv("tomato.csv")

correlation=cor(data[2:7], method = "pearson")

test=rcorr(as.matrix(data[2:7], type="pearson"))
test
```

**Result in console**

```
Console    Terminal    Jobs
R 4.0.2 · D:/mydata/
> test=rcorr(as.matrix(data[2:7], type="pearson"))
> test
            PH   NPB   CHL Phenol Moisture    DS
PH        1.00  0.54 -0.04  -0.03    -0.09  0.00
NPB       0.54  1.00 -0.05  -0.01    -0.04 -0.03
CHL      -0.04 -0.05  1.00   0.12    -0.03 -0.09
Phenol   -0.03 -0.01  0.12   1.00     0.00 -0.82
Moisture -0.09 -0.04 -0.03   0.00     1.00 -0.06
DS        0.00 -0.03 -0.09  -0.82    -0.06  1.00

n= 235


P
         PH     NPB    CHL    Phenol Moisture DS
PH              0.0000 0.5814 0.6388 0.1471   0.9742
NPB      0.0000        0.4155 0.8531 0.5406   0.6805
CHL      0.5814 0.4155        0.0610 0.6244   0.1649
Phenol   0.6388 0.8531 0.0610        0.9687   0.0000
Moisture 0.1471 0.5406 0.6244 0.9687          0.3679
DS       0.9742 0.6805 0.1649 0.0000 0.3679
>
```

The "rcorr()" function in R generates an output that is a list containing several key elements:

- "*r*": This is the correlation matrix.
- "*n*": This matrix shows the number of observations used for each pair of variables in the analysis.
- "p": These are the *p*-values corresponding to the significance levels of the correlations.

To export the *p*-values of the correlation matrix to a .doc file, the "bruceR" package can be utilized

by using this code:

```
print_table(test$P, file="test of significance")
```

Furthermore, for those seeking to obtain the output of correlation values along with their significant *p*-values in a column format, the "correlation" package in R offers a solution. After installing and launching this package, the following code can be used:

```
correlation(data, include_factors = TRUE, method = "auto")
```

**In script editor**

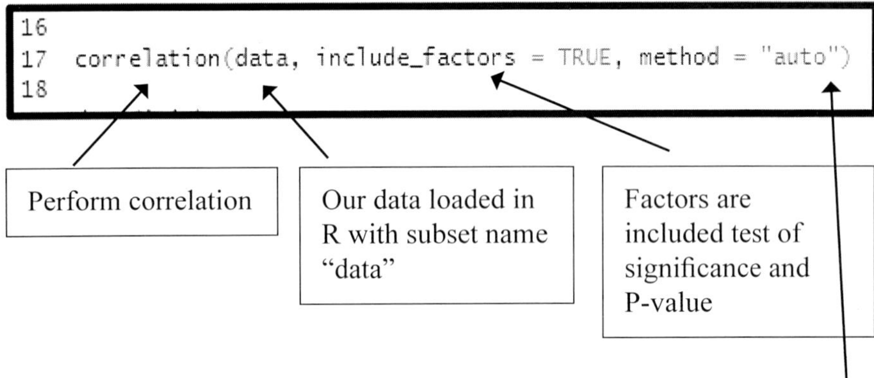


A character string indicating which correlation coefficient is to be used for the test. One of "Pearson" (default), "Kendall," "spearman" (but see also the robust argument), "biserial," "polychoric," "tetrachoric," "biweight," "distance," "percentage" (for percentage bend correlation), "Blomqvist" (for Blomqvist's coefficient), "Hoeffding" (for Hoeffding's D), "gamma," "Gaussian" (for Gaussian Rank correlation) or "Shepherd" (for Shepherd's Pi correlation). Setting "auto" will attempt at selecting the most relevant method (polychoric when ordinal factors involved, tetrachoric when dichotomous factors involved, point-biserial if one dichotomous and one continuous and pearson otherwise).

## Output in console

```
Console   Terminal   Jobs
R 4.0.2 · D:/mydata/
> correlation(data, include_factors = TRUE, method = "auto")
# Correlation Matrix (auto-method)

Parameter1 | Parameter2 |     r |          95% CI |  t(30) |         p
-----------------------------------------------------------------------
mpg        |       disp | -0.85 | [-0.92, -0.71] |  -8.75 | < .001***
mpg        |         hp | -0.78 | [-0.89, -0.59] |  -6.74 | < .001***
mpg        |       drat |  0.68 | [ 0.44,  0.83] |   5.10 | < .001***
mpg        |         wt | -0.87 | [-0.93, -0.74] |  -9.56 | < .001***
mpg        |       qsec |  0.42 | [ 0.08,  0.67] |   2.53 | 0.053
disp       |         hp |  0.79 | [ 0.61,  0.89] |   7.08 | < .001***
disp       |       drat | -0.71 | [-0.85, -0.48] |  -5.53 | < .001***
disp       |         wt |  0.89 | [ 0.78,  0.94] |  10.58 | < .001***
disp       |       qsec | -0.43 | [-0.68, -0.10] |  -2.64 | 0.053
hp         |       drat | -0.45 | [-0.69, -0.12] |  -2.75 | 0.050*
hp         |         wt |  0.66 | [ 0.40,  0.82] |   4.80 | < .001***
hp         |       qsec | -0.71 | [-0.85, -0.48] |  -5.49 | < .001***
drat       |         wt | -0.71 | [-0.85, -0.48] |  -5.56 | < .001***
drat       |       qsec |  0.09 | [-0.27,  0.43] |   0.50 | 0.678
wt         |       qsec | -0.17 | [-0.49,  0.19] |  -0.97 | 0.678

p-value adjustment method: Holm (1979)
Observations: 32
>
```

The correlation analysis results can be conveniently saved in a .doc format using the "bruceR" package, as detailed in this chapter. To achieve this, execute the following code, referencing the name of the subset containing the correlation results:

```
correlation_table=correlation(data, include_factors = TRUE, method = "auto")
```

To save the subset named "correlation_table" in a .doc format, execute the appropriate code using the "bruceR" package. This will result in the "correlation_table" data being exported as a well-formatted table in a Word document.

| | Parameter1 | Parameter2 | r | CI | CI_low | CI_high | t | df_error | p | Method | n_Obs |
|---|---|---|---|---|---|---|---|---|---|---|---|
| 1 | mpg | disp | −0.848 | 0.950 | −0.923 | −0.708 | −8.747 | 30.000 | 0.000 | Pearson correlation | 32.000 |
| 2 | mpg | hp | −0.776 | 0.950 | −0.885 | −0.586 | −6.742 | 30.000 | 0.000 | Pearson correlation | 32.000 |
| 3 | mpg | drat | 0.681 | 0.950 | 0.436 | 0.832 | 5.096 | 30.000 | 0.000 | Pearson correlation | 32.000 |
| 4 | mpg | wt | −0.868 | 0.950 | −0.934 | −0.744 | −9.559 | 30.000 | 0.000 | Pearson correlation | 32.000 |
| 5 | mpg | qsec | 0.419 | 0.950 | 0.082 | 0.670 | 2.525 | 30.000 | 0.053 | Pearson correlation | 32.000 |
| 6 | disp | hp | 0.791 | 0.950 | 0.611 | 0.893 | 7.080 | 30.000 | 0.000 | Pearson correlation | 32.000 |
| 7 | disp | drat | −0.710 | 0.950 | −0.849 | −0.481 | −5.526 | 30.000 | 0.000 | Pearson correlation | 32.000 |
| 8 | disp | wt | 0.888 | 0.950 | 0.781 | 0.944 | 10.576 | 30.000 | 0.000 | Pearson correlation | 32.000 |
| 9 | disp | qsec | −0.434 | 0.950 | −0.680 | −0.100 | −2.636 | 30.000 | 0.053 | Pearson correlation | 32.000 |
| 10 | hp | drat | −0.449 | 0.950 | −0.690 | −0.119 | −2.750 | 30.000 | 0.050 | Pearson correlation | 32.000 |
| 11 | hp | wt | 0.659 | 0.950 | 0.403 | 0.819 | 4.796 | 30.000 | 0.000 | Pearson correlation | 32.000 |
| 12 | hp | qsec | −0.708 | 0.950 | −0.848 | −0.477 | −5.495 | 30.000 | 0.000 | Pearson correlation | 32.000 |
| 13 | drat | wt | −0.712 | 0.950 | −0.850 | −0.484 | −5.561 | 30.000 | 0.000 | Pearson correlation | 32.000 |
| 14 | drat | qsec | 0.091 | 0.950 | −0.266 | 0.426 | 0.502 | 30.000 | 0.678 | Pearson correlation | 32.000 |
| 15 | wt | qsec | −0.175 | 0.950 | −0.493 | 0.185 | −0.972 | 30.000 | 0.678 | Pearson correlation | 32.000 |

## Visualize Correlation Matrix

In R, visualizing a correlation matrix can be achieved through various methods, including the "**corrplot()**" function, scatter plots, and heat maps. The "**corrplot()**" function, part of the package with the same name, is particularly effective for creating a correlogram—a graphical representation of the correlation matrix. It highlights the most correlated variables and allows for customization of the display.

Key points of the "**corrplot()**" function include:

Color-coding: Correlation coefficients are visually distinguished by color, making it easy to identify the strength of associations.

Reordering: The correlation matrix can be reordered based on the degree of association between variables, enhancing the clarity of the relationships.

To visualize a correlation matrix using "corrplot()," follow these steps:

i. **Install and activate the package:**

```
install.packages("corrplot")

library(corrplot)
```

ii. **Prepare and execute the data:**

Use a dataset with varied correlation values for a clearer understanding of the graphical output.

```
corrplot(res, method = "number")
```

Note:

- corrplot: Function to plot the correlation matrix.
- res: The subset or the specific correlation matrix to visualize.
- method = "number:" The chosen method for the correlation plot, in this case, displays the correlation coefficients numerically.

**Description of "corrplot" function**

- **corr:** This parameter specifies the correlation matrix to be visualized. The matrix should be square unless the "original" order is used. For matrices that aren't correlation matrices, set "is.corr = FALSE" for conversion.
- **Method:** This argument determines the visualization style of the correlation matrix. The function supports several methods, including "circle" (the default), "square," "ellipse," "number," "pie," "shade," and "color." Each method offers a unique visual representation of the correlation data.

  The size of the circles or squares in the "circle" and "square" methods represent the absolute value of the correlation coefficients. The "pie" and "shade" methods were adapted from Michael Friendly's work, incorporating adjustments to shading, and the "ellipse" method was derived from the work of D.J. Murdoch and E.D. Chow.
- **Type:** This parameter determines the portion of the matrix to display. Options include "full" (the default setting), which shows the entire matrix, "upper" for the upper triangular part, and "lower" for the lower triangular part.
- **add:** A logical argument. If set to TRUE, the correlation plot is added onto an existing plot. If FALSE, a new plot is created for the visualization.
- **col:** This is a vector specifying the colors of the graphical elements (glyphs) in the plot. The colors are distributed uniformly across the range specified in the "col.lim" interval. If "col" is set to NULL, the function uses a default color palette, which can be seen in the documentation for the "colorRampPalette(col2)(200)" function.
- **col.lim:** This parameter sets the limits for the color range, defined as an interval ($x1$, $x2$). When "col.lim" is set to NULL and "is.corr" is TRUE, the default range is set between −1 and 1. If "is.corr" is FALSE, "col.

lim" defaults to the range between the minimum and maximum values of the correlation matrix.

Note: when col.lim is specified while is.corr is TRUE, the method for assigning colors remains uniformly distributed across the −1 to 1 range.

- **bg:** Sets the background color of the plot.
- **title:** A character string that specifies the title of the graph.
- **is.corr:** A logical value indicating whether the input matrix is a correlation matrix. Set is.corr = FALSE to visualize matrices that are not correlation matrices.
- **diag:** A logical parameter that determines whether to display the correlation coefficients on the principal diagonal of the matrix.
- **outline:** This can be a logical value or a character string. It controls whether to plot an outline around the shapes (circles, squares, ellipses) used in the graph. For the "pie" method, it specifies the color of the circle outlining the pie. When set to TRUE, the default outline color is "black."
- **addgrid.col:** Specifies the color of the grid in the plot.
- **addCoef.col:** Determines the color of the correlation coefficients displayed on the graph.
- **addCoefasPercent:** A logical parameter that, if set to TRUE, converts the correlation coefficients into a percentage format, which can be useful for saving space in the plot.
- **order:** This character parameter sets the ordering method of the correlation matrix. Options include:
  - ✓ "original": Maintains the original order of the matrix (default setting).
  - ✓ "AOE": Orders based on the angular order of the eigenvectors.
  - ✓ "FPC": Uses the first principal component order.
  - ✓ "hclust": Applies hierarchical clustering order.
  - ✓ "alphabet": Arranges variables in alphabetical order.
- **addrect:** An integer specifying how many rectangles to draw on the graph, based on hierarchical clustering. It's effective only when

order = "hclust." The default is NULL, which means no rectangles are drawn.

- **rect.col:** Sets the color of the rectangle borders. Applicable when addrect is 1 or more.
- **rect.lwd:** Defines the line width of the rectangle borders. Relevant when addrect is 1 or more.
- **tl.pos:** Determines the position of text labels with options like "lt" (left and top), "ld" (left and diagonal), "td" (top and diagonal), "d" (diagonal), or "n" (no text label). The default varies based on the plot type.
- **tl.cex:** Numeric value for the size of text labels (variable names).
- **tl.col:** Color of the text labels.
- **tl.offset:** Numeric value for offsetting text labels.
- **tl.srt:** Numeric value for the degree of rotation of text labels.
- **cl.pos:** Sets the position of the color legend, with options "r" (right), "b" (bottom), or "n" (no legend).
- **cl.length:** Integer specifies the number of text labels in the color legend.
- **cl.cex:** Numeric value for the size of number labels in the color legend.
- **cl.ratio:** Numeric value to adjust the width of the color legend.
- **cl.align.text:** Sets alignment ("l" for left, "c" for center, "r" for right) of number labels in the color legend.
- **cl.offset:** Numeric value for offsetting number labels in the color legend.
- **number.cex:** Size parameter for correlation coefficients text in the plot.
- **number.font:** Font parameter for correlation coefficients text in the plot.
- **number.digits:** Number of decimal digits for correlation coefficients in the plot.
- **addshade:** Specifies the shading style ("negative," "positive," or "all") for the "shade" method.
- **shade.lwd:** Line width for the shading.
- **shade.col:** Color of the shading lines.
- **p.mat:** Matrix of $p$-values for significance testing.

- **sig.level:** Significance level for considering correlation coefficients as significant.
- **Insig:** Method for marking insignificant correlation coefficients.
- **pch:** Character to use for marking insignificant coefficients.
- **pch.col:** Color of the character marking insignificant coefficients.
- **pch.cex:** Size of the character marking insignificant coefficients.
- **plotCI:** Method for plotting confidence intervals.
- **lowCI.mat:** Matrix of the lower bounds of confidence intervals.
- **uppCI.mat:** Matrix of the upper bounds of confidence intervals.
- **na.label:** Label for rendering NA cells, default is "?."
- **na.label.col:** Color for rendering NA cells.
- **win.asp:** Aspect ratio for the entire plot, compatible with "circle" and "square" methods.

## Different Types of Correlation Plots or Correlogram

| Code 1 | `corrplot(res, method = 'number')` |
|---|---|
| output | 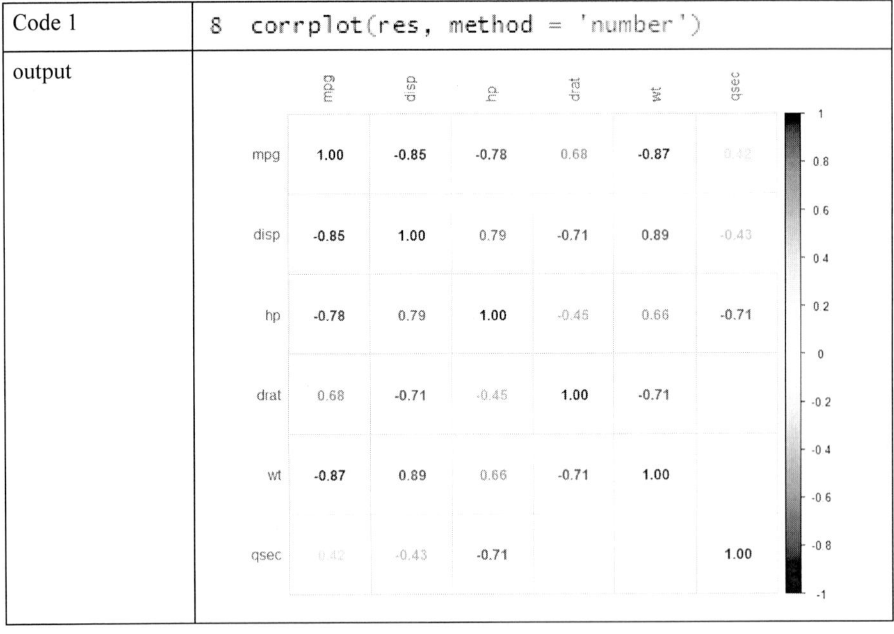 |

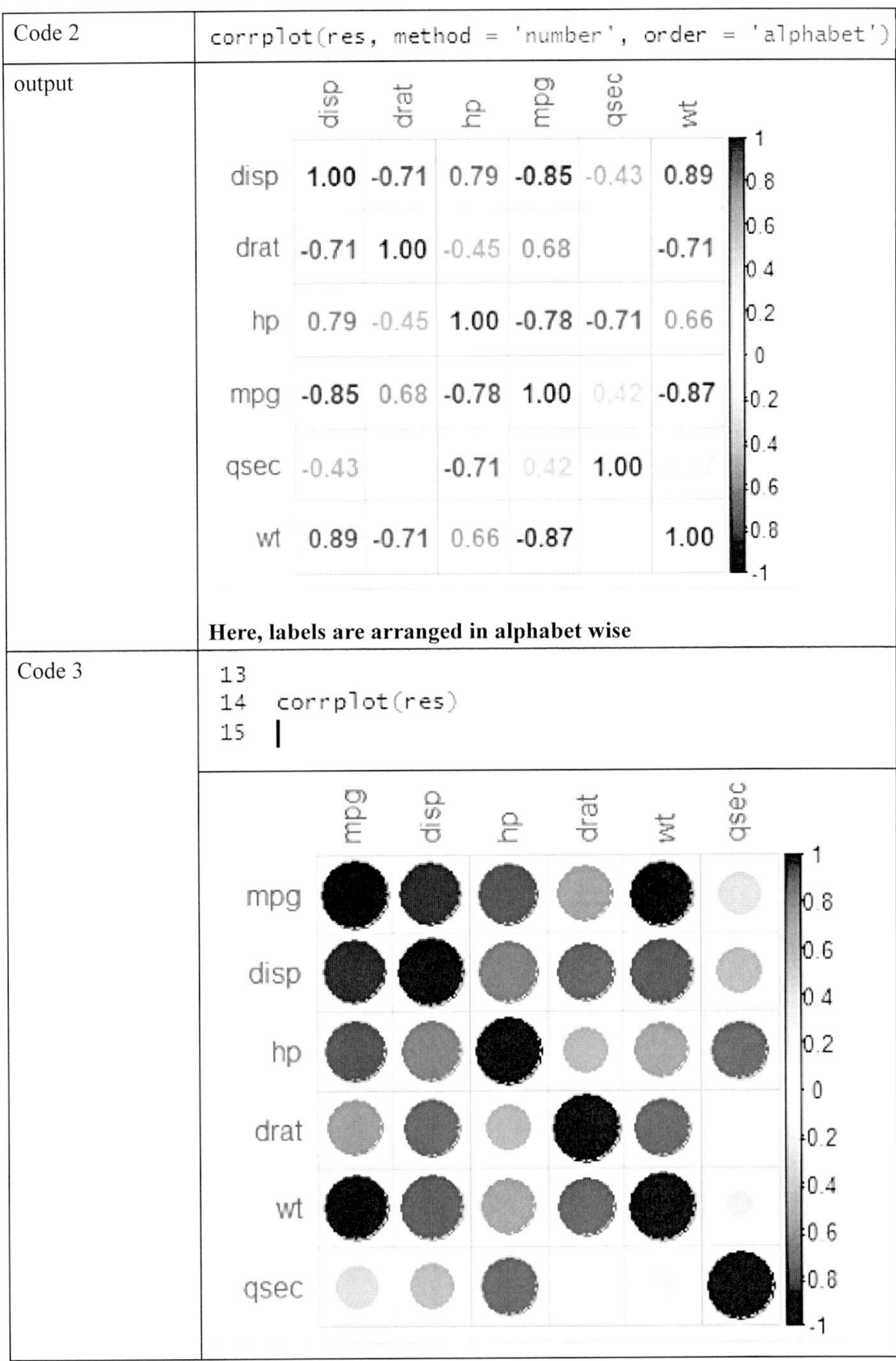


| | |
|---|---|
| Code 2 | `corrplot(res, method = 'number', order = 'alphabet')` |
| output | **Here, labels are arranged in alphabet wise** |
| Code 3 | `corrplot(res)` |

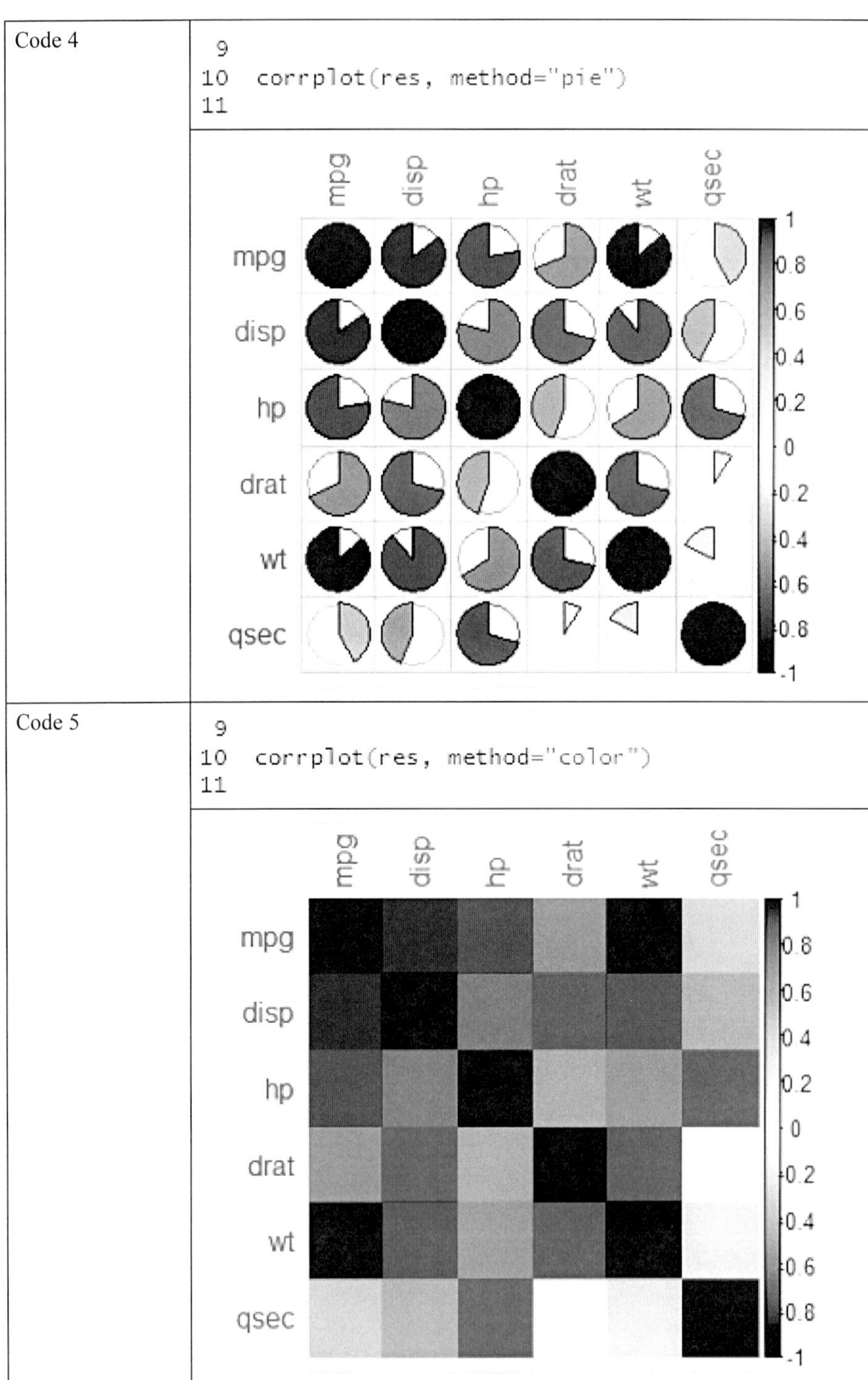


Code 4

```
corrplot(res, method="pie")
```

Code 5

```
corrplot(res, method="color")
```

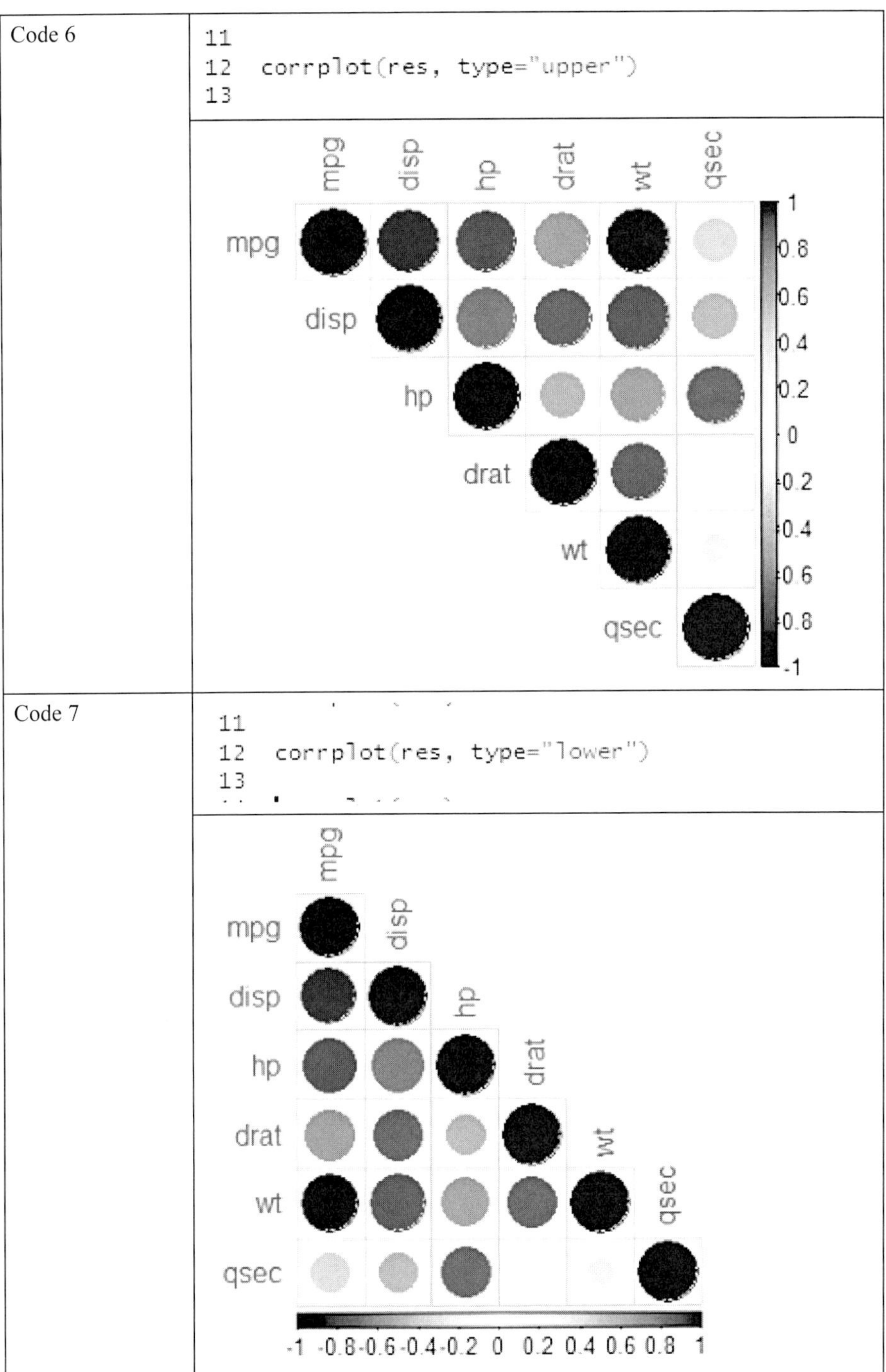
Code 6
corrplot(res, type="upper")
mpg
disp
hp
drat
wt
qsec
1
0.8
0.6
0.4
0.2
0
-0.2
-0.4
-0.6
-0.8
-1
Code 7
corrplot(res, type="lower")
mpg
disp
hp
drat
wt
qsec
-1 -0.8 -0.6 -0.4 -0.2 0 0.2 0.4 0.6 0.8 1

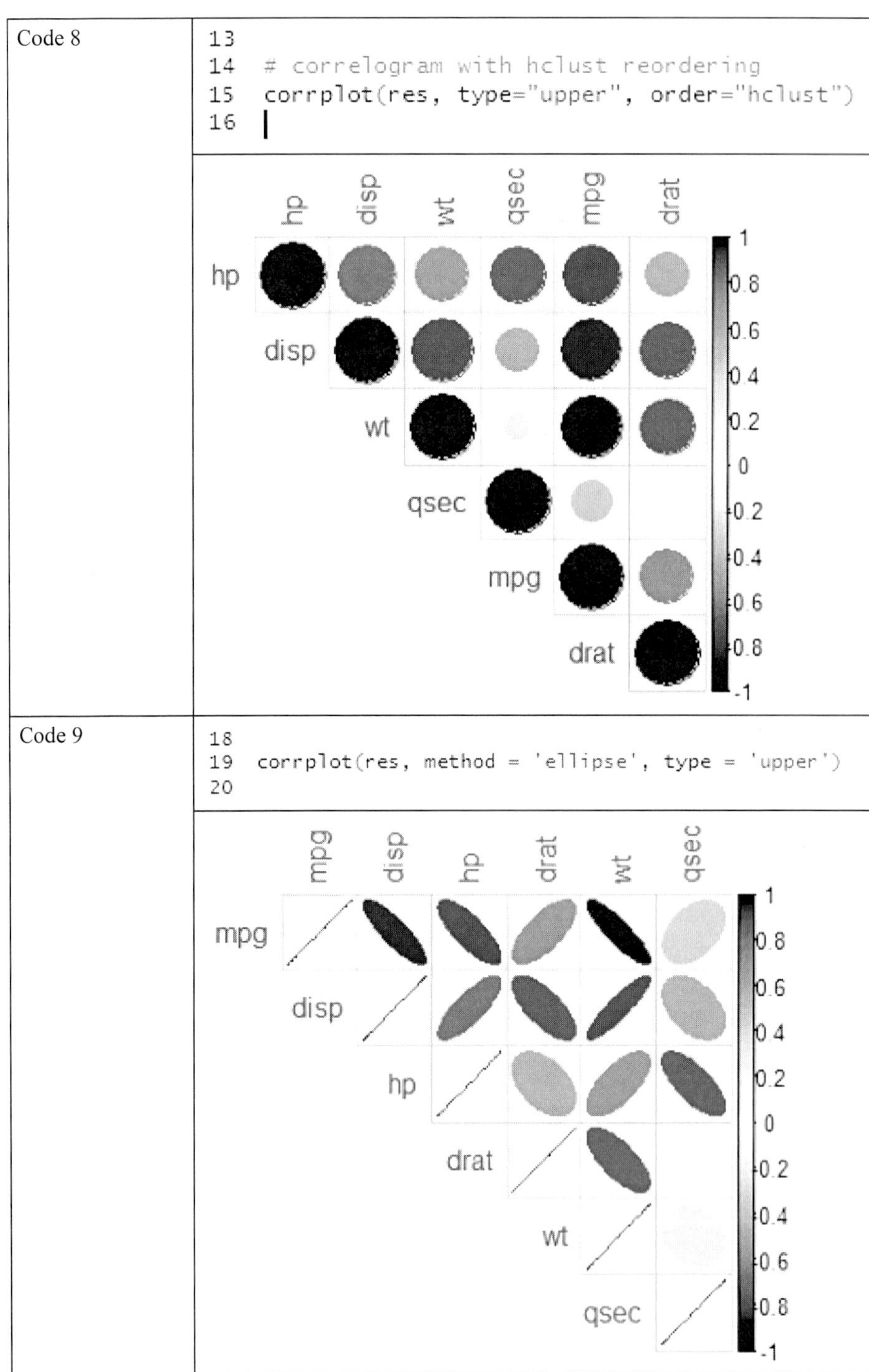

Code 8

```
# correlogram with hclust reordering
corrplot(res, type="upper", order="hclust")
```



Code 9

```
corrplot(res, method = 'ellipse', type = 'upper')
```

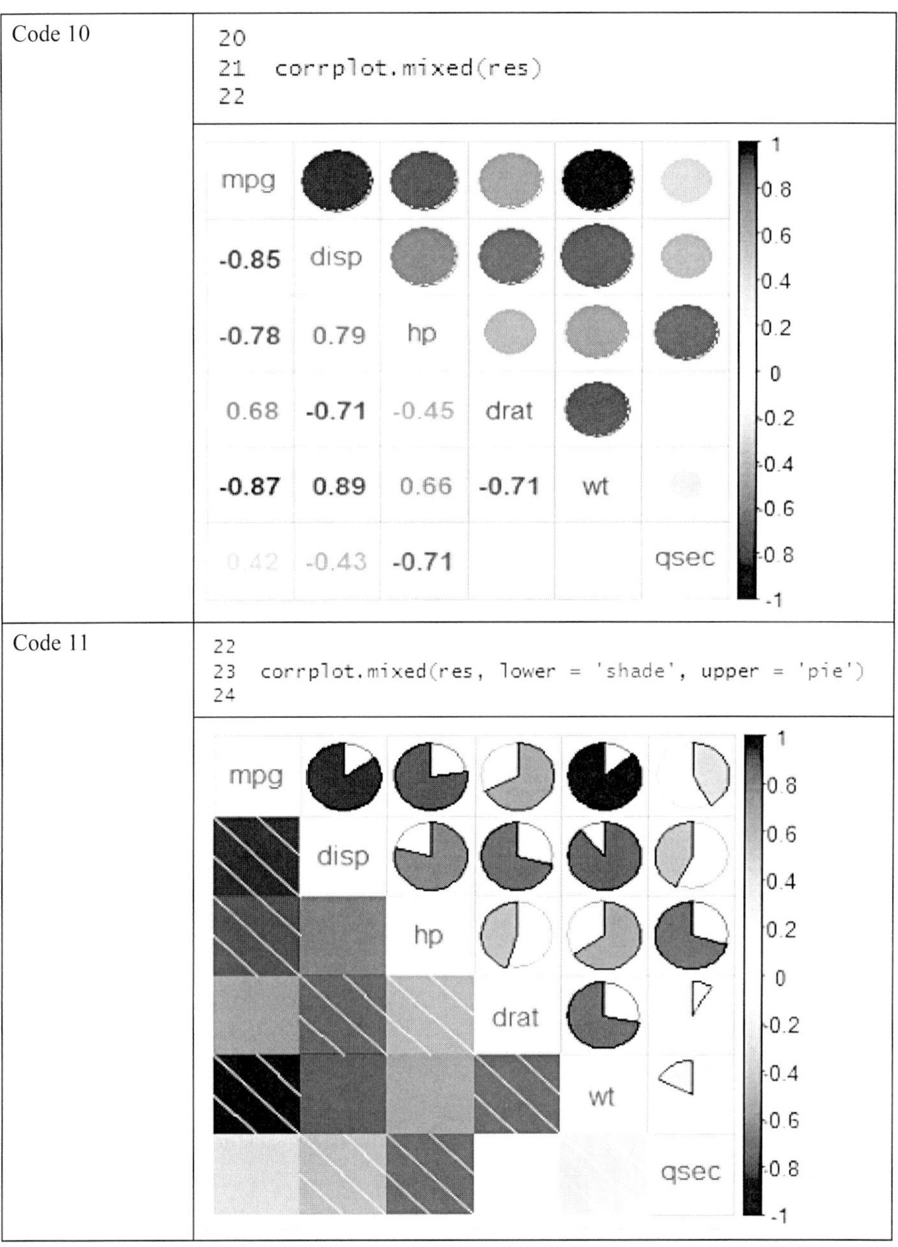
Code 10
corrplot.mixed(res)
mpg
disp
hp
drat
wt
qsec
-0.85
-0.78 0.79
0.68 -0.71 -0.45
-0.87 0.89 0.66 -0.71
0.42 -0.43 -0.71
1
0.8
0.6
0.4
0.2
0
-0.2
-0.4
-0.6
-0.8
-1
Code 11
corrplot.mixed(res, lower = 'shade', upper = 'pie')
mpg
disp
hp
drat
wt
qsec
1
0.8
0.6
0.4
0.2
0
-0.2
-0.4
-0.6
-0.8
-1

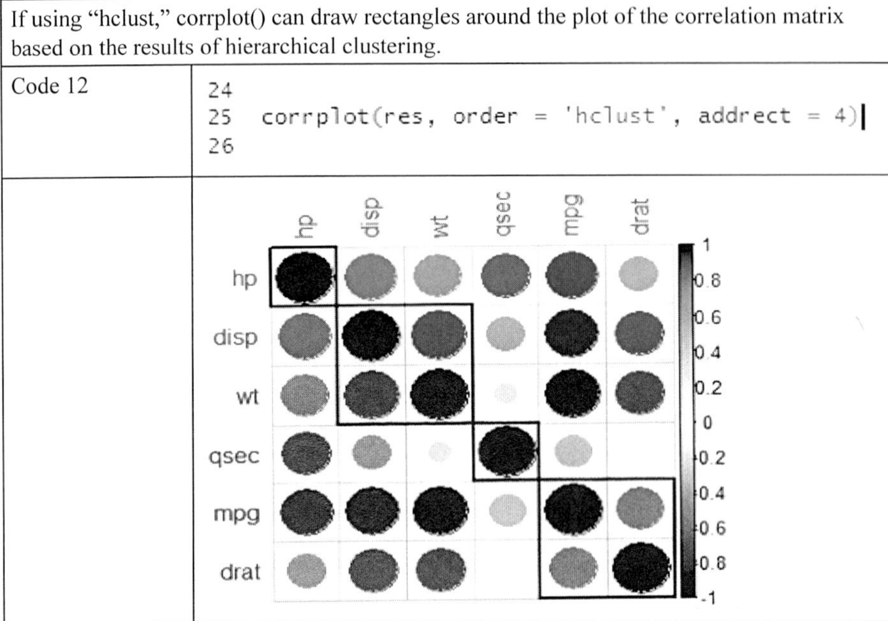

| If using "hclust," corrplot() can draw rectangles around the plot of the correlation matrix based on the results of hierarchical clustering. | |
|---|---|
| Code 12 | `corrplot(res, order = 'hclust', addrect = 4)` |
| | |



To achieve a comprehensive single plot that includes the distribution of each variable on the diagonal, bivariate scatter plots with fitted lines off-diagonal, and correlation values with significance levels denoted by stars, you'll need to use the "PerformanceAnalytics" package in R. This package allows for the creation of such detailed charts, where the significance levels are indicated by different symbols corresponding to various p-value thresholds.

To use "PerformanceAnalytics" for this purpose, you would need to install and load the package with the following R commands:

**install.packages("PerformanceAnalytics")**

**library(PerformanceAnalytics)**

**Download and launch it then**

**Run the code**

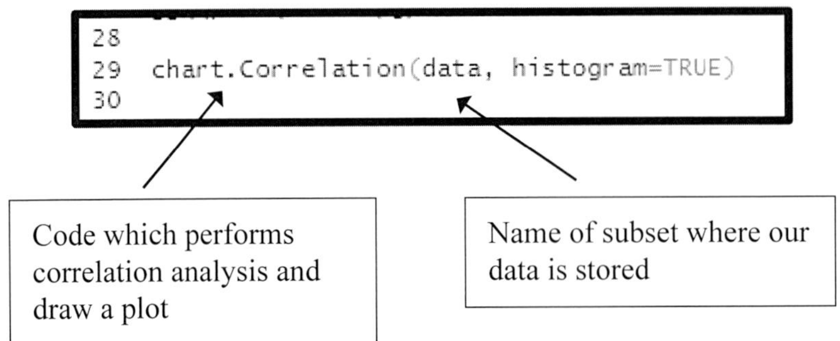

## Output in the plot area

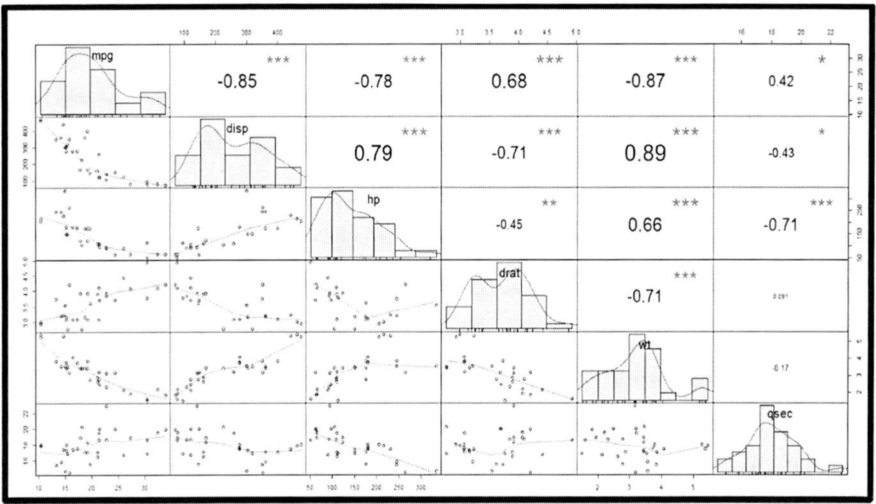

# 6

# Path Analysis

This chapter provides an insight into the genotypic and phenotypic correlations between plant traits. It incorporates the study of genotypic and phenotypic path analysis and the understanding of genetic variability, all illustrated through practical examples in R, relevant to plant breeding research.

## Path Analysis

Path analysis explains not only the direct impact of one character on another but also indirect effects mediated through other variables in the system. The approach is similar to the analysis of variance and is sometimes referred to as the analysis of correlation coefficients. A notable feature of path coefficients is that they are standardized partial regression coefficients, devoid of units. This standardization allows for the direct effects of different variables to be compared and ranked based on their magnitude. This approach was first applied in plant studies by Dewey and Lu in 1959, particularly to examine the factors influencing seed yield.

## Path Coefficients

In the computation of path coefficients, a path diagram is translated into a series of simultaneous equations. These equations reveal the direct and indirect contributions of causal variables to the outcome. For example, in the field of plant breeding: consider the yield of forage (Y) being influenced by different variables. These could include the leaf length (X1), leaf width (X2), and number of leaves per plant (X3). The path diagram for these variables can be given as follows:

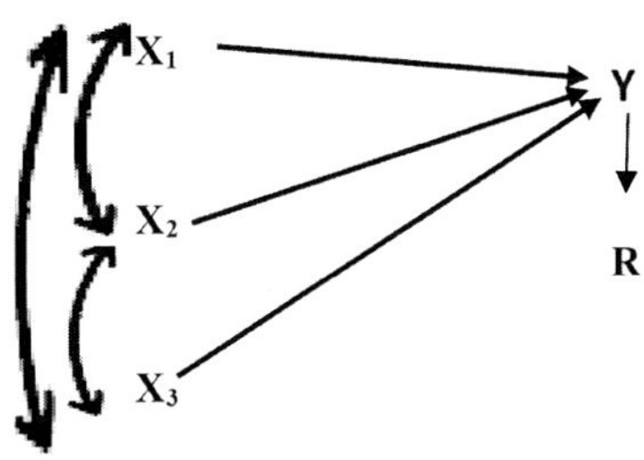

Path coefficient simultaneous equations can be formulated as follows:

$$r_{yx_1} = P_{1Y} + P_{2Y} * r_{x_1x_2} + P_{3Y} * r_{x_1x_2}, r_{yx_2} = P_{1Y} * r_{x_1x_2} + P_{2Y} + P_{3Y} * r_{x_2x_3}$$

$$r_{yx_2} = P_{1Y} * r_{x_1x_2} + P_{2Y} + P_{3Y} * r_{x_2x_3},$$

$$r_{yx_3} = P_{1Y} * r_{x_1x_3} + P_{2Y} * r_{x_2x_3} + P_{3Y}.$$

The data can be structured into a matrix format as follows:

$$\begin{pmatrix} r_{yx_1} \\ r_{yx_2} \\ r_{yx_3} \end{pmatrix} = \begin{pmatrix} 1 & r_{x_1x_2} & r_{x_1x_3} \\ r_{x_1x_2} & 1 & r_{x_2x_3} \\ r_{x_1x_3} & r_{x_2x_3} & 1 \end{pmatrix} \begin{pmatrix} P_{1Y} \\ P_{2Y} \\ P_{3Y} \end{pmatrix}.$$

1[st] matrix represents the correlation coefficients between these component characters and yield. The 2[nd] matrix denotes the correlation coefficients correlation coefficients among characters. The objective is to determine the $P_{iy}$ values, or path coefficient values.

## Genotypic and Phenotypic correlation in R

### *Genotypic and Phenotypic Correlation Analysis of Replicated Data*

Genotypic and phenotypic analysis stands as acritical statistical method in plant breeding, necessitating the use of the "variability" package. This analysis focuses on replicated data of variables, each organized in a unique arrangement, to effectively evaluate both genotypic and phenotypic correlations.

Below screenshots shows how to arrange data in Excel format.

| GEN | REP | TA | SI | PH | EH | EPP | EL | KPR | EG | KRP | NKE | Pro | TSS | Car | GCY |
|---|---|---|---|---|---|---|---|---|---|---|---|---|---|---|---|
| 1 | 1 | 69 | 77 | 119 | 51.37 | 1 | 12.66 | 31 | 12.67 | 13.33 | 413.33 | 16.3 | 2.3 | 2.35 | 65 |
| 2 | 1 | 65 | 72 | 109.2 | 29.43 | 1 | 14.2 | 24 | 10.75 | 14 | 336 | 17.21 | 1.6 | 2 | 62.5 |
| 3 | 1 | 70 | 76 | 124.2 | 50.13 | 1 | 13.5 | 35.3 | 11.83 | 15.36 | 542.21 | 16.26 | 1.7 | 6.35 | 50 |
| 4 | 1 | 71 | 79 | 141.6 | 83.15 | 1 | 17.3 | 32.65 | 14 | 11.53 | 376.45 | 15.9 | 1.85 | 2.37 | 136.67 |
| 5 | 1 | 71 | 77 | 158 | 70.15 | 1 | 17.5 | 35.36 | 13.25 | 14 | 495.04 | 12.6 | 1 | 4.63 | 100 |
| 6 | 1 | 70 | 75 | 95.4 | 25.84 | 1.67 | 10.33 | 29.65 | 12.33 | 14 | 415.1 | 13.02 | 2.4 | 9.33 | 69.44 |
| 7 | 1 | 61 | 67 | 90.2 | 62.3 | 2 | 13.67 | 29.33 | 12.33 | 13.33 | 391.11 | 13.62 | 2.36 | 3.65 | 150 |
| 8 | 1 | 75 | 80 | 86.6 | 42.38 | 1.5 | 15 | 23.2 | 13 | 12.36 | 286.75 | 15.26 | 1.75 | 3.14 | 44.74 |
| 9 | 1 | 73 | 80 | 129.8 | 35.81 | 1.67 | 15.5 | 33 | 10 | 18 | 594 | 14.32 | 2.63 | 7.32 | 125 |
| 10 | 1 | 70 | 77 | 81.28 | 24.33 | 1 | 15.64 | 32.45 | 11 | 13 | 421.85 | 13.62 | 1.7 | 5.33 | 25 |

| | | | | | | | | | | | | | | | | |
|---|---|---|---|---|---|---|---|---|---|---|---|---|---|---|---|---|
| 44 | 43 | 1 | 71 | 79 | 112.6 | 43.33 | 1 | 13.69 | 27 | 13.56 | 15 | 405 | 15.86 | 2.5 | 10.35 | 73 |
| 45 | 44 | 1 | 72 | 77 | 136.2 | 35.71 | 1.33 | 15 | 31.67 | 12.67 | 14 | 443.33 | 14.7 | 2.6 | 8.37 | 66.67 |
| 46 | 45 | 1 | 73 | 77 | 139.8 | 61.67 | 1 | 12.25 | 32 | 15 | 15.33 | 490.67 | 15.33 | 2.37 | 7.33 | 94.21 |
| 47 | 1 | 2 | 73 | 80 | 108.6 | 47.91 | 1.33 | 11.5 | 28.65 | 11.75 | 15 | 429.75 | 14.32 | 2.33 | 4.33 | 55.42 |
| 48 | 2 | 2 | 64 | 70 | 116.3 | 36.14 | 2 | 14.83 | 25 | 9 | 12 | 300 | 16.32 | 1.64 | 2.37 | 50 |
| 49 | 3 | 2 | 68 | 74 | 115.4 | 53.31 | 2 | 17.86 | 38 | 15 | 17.33 | 658.67 | 16.35 | 1.75 | 6.35 | 50 |
| 50 | 4 | 2 | 70 | 77 | 134.3 | 70.9 | 1 | 18 | 35.33 | 15 | 12.67 | 447.56 | 14.85 | 1.9 | 1.23 | 158.33 |
| 51 | 5 | 2 | 73 | 80 | 145.36 | 65.23 | 1 | 16.33 | 39 | 13.83 | 13.33 | 520 | 11.42 | 0.99 | 3.25 | 100 |
| 52 | 6 | 2 | 72 | 78 | 85.47 | 19.54 | 1.69 | 14.17 | 30.67 | 14 | 16.67 | 511.11 | 11.87 | 2.48 | 7.53 | 105.63 |
| 53 | 7 | 2 | 60 | 69 | 100.23 | 55.9 | 1.66 | 16.17 | 32.33 | 13.33 | 12.67 | 409.56 | 12.82 | 2.16 | 6.33 | 179.83 |
| 54 | 8 | 2 | 78 | 85 | 80.6 | 38.9 | 1.5 | 14.56 | 22 | 12 | 11.56 | 254.32 | 16.21 | 1.72 | 3.15 | 37.5 |
| 55 | 9 | 2 | 74 | 80 | 138.6 | 42.81 | 1.33 | 15.83 | 31 | 14.33 | 16.36 | 507.16 | 13.42 | 1.92 | 6.33 | 155.17 |
| 56 | 10 | 2 | 70 | 76 | 126.48 | 30.93 | 1 | 17.33 | 35.33 | 13.17 | 14.67 | 518.22 | 12.25 | 1.7 | 5.35 | 116.67 |

| | | | | | | | | | | | | | | | | |
|---|---|---|---|---|---|---|---|---|---|---|---|---|---|---|---|---|
| 86 | 40 | 2 | 74 | 80 | 84.92 | 23.56 | 1.66 | 12.33 | 26 | 11.83 | 14 | 364 | 14.29 | 2.21 | 16.66 | 69.17 |
| 87 | 41 | 2 | 72 | 77 | 146.5 | 38.48 | 1.66 | 11.33 | 28.67 | 11 | 12.67 | 363.11 | 14.33 | 2.05 | 5.63 | 83 |
| 88 | 42 | 2 | 74 | 81 | 100.76 | 32.18 | 1.33 | 13.33 | 27.67 | 13.33 | 14 | 387.33 | 12.05 | 1.43 | 14.26 | 110.83 |
| 89 | 43 | 2 | 74 | 81 | 108.5 | 40.78 | 1 | 14.33 | 29 | 12 | 14 | 406 | 12.34 | 2.64 | 9.36 | 75 |
| 90 | 44 | 2 | 71 | 79 | 126.45 | 32.54 | 1.66 | 14.33 | 28.64 | 11 | 13.33 | 381.87 | 14.28 | 2.62 | 6.58 | 83 |
| 91 | 45 | 2 | 75 | 79 | 123.75 | 55.47 | 1.66 | 13.75 | 27 | 14.25 | 16 | 432 | 14.45 | 2.34 | 8.45 | 103.75 |
| 92 | | | | | | | | | | | | | | | | |

The dataset for this analysis is stored in a .csv file, organized as follows:

- **Column A (GEN):** Contains the treatment vector or the names of the varieties.
- **Column B (REP):** Holds the replication vector or the number of replication.
- **Columns C to P:** Represent various variables (from the 3rd to the 16th column).

In this dataset, green cob yield (GCY) is the dependent variable and is placed in the last column. This is important for path analysis, as the dependent variable needs to be excluded from the set of independent variables (columns 3–15).

This same dataset is utilized across various analyses including genotypic correlation analysis, phenotypic correlation analysis, genotypic path analysis, phenotypic path analysis, genetic variability analysis, and covariance analysis.

To proceed, save the file in .csv format and read it into RStudio using an appropriately named data subset. Then, download and activate the "variability" package for the analysis.

### *Genotypic Correlation Analysis with 5% and 1% Level of Significance*

*Code for genotypic correlation analysis*

Code: `g_corr=geno.corr(data[3:16],data$GEN,data$REP)`

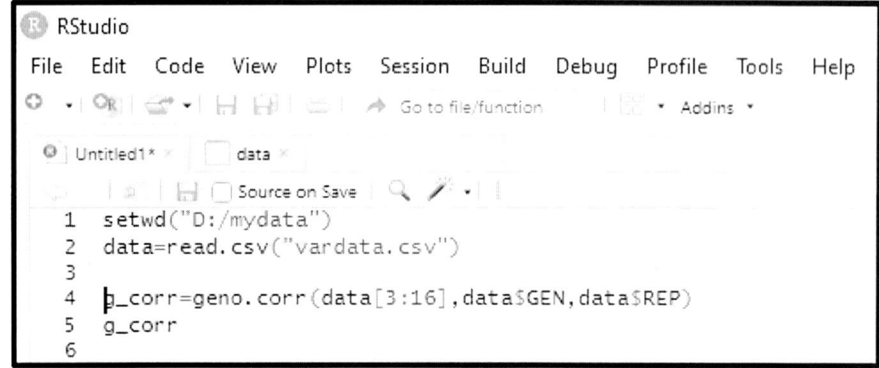


**Description of code:**

- **g_corr:** name of subset where analyzed genotypic correlation is saved
- **geno.corr:** name of function
- **data[3:16]:** name of data with address of variable in column
- **data$GEN:** treatment or genotypic vactor
- **data$REP:** replication vector available in data sheet

Now, execute the specified code in RStudio and observe the results displayed in the console.

```
Console  Terminal  Jobs
R 4.0.2 · D:/mydata/
> g_corr=geno.corr(data[3:16],data$GEN,data$REP)
> g_corr
$GenotypicCorrelation
    TA          SI          PH          EH          EPP         EL          KPR
TA  1 **        0.9832 **   -0.1452 NS -0.1719 NS -0.457 **   -0.2527 NS -0.1264 NS
SI  0.9832 **   1 **        -0.2206 NS -0.2007 NS -0.4949 **  -0.2252 NS -0.1045 NS
PH  -0.1452 NS -0.2206 NS 1 **        0.6776 **   -0.2211 NS 0.248 NS    0.3864 **
EH  -0.1719 NS -0.2007 NS 0.6776 **   1 **        -0.3312 *   0.2257 NS   0.3117 *
EPP -0.457 **   -0.4949 **  -0.2211 NS -0.3312 *   1 **        -0.1659 NS -0.1957 NS
EL  -0.2527 NS -0.2252 NS 0.248 NS    0.2257 NS   -0.1659 NS 1 **        0.7161 **
KPR -0.1264 NS -0.1045 NS 0.3864 **   0.3117 *    -0.1957 NS 0.7161 **   1 **
EG  -0.3224 *   -0.3198 *   0.3087 *    0.4782 **   -0.0261 NS 0.4345 **   0.4906 **
KRP -0.1007 NS -0.1315 NS 0.2188 NS   0.1509 NS   0.0888 NS   -0.1946 NS 0.1838 NS
NKE -0.1542 NS -0.1518 NS 0.3819 **   0.3024 *    -0.1189 NS 0.5402 **   0.9239 **
Pro 0.198 NS    0.1952 NS   -0.0202 NS 0.1918 NS   -0.1771 NS -0.1199 NS 0.0063 NS
TSS -0.2773 NS -0.2591 NS 0.0301 NS   -0.0842 NS 0.2856 NS   -0.3229 *   -0.2117 NS
Car 0.0992 NS   0.1094 NS   -0.0322 NS -0.3159 *   0.0775 NS   0.0623 NS   0.1905 NS
GCY -0.4938 **  -0.4662 **  0.1881 NS   0.3081 *    0.3777 *    0.59 **     0.5012 **
    EG          KRP         NKE         Pro         TSS         Car         GCY
TA  -0.3224 *   -0.1007 NS -0.1542 NS 0.198 NS    -0.2773 NS 0.0992 NS   -0.4938 **
SI  -0.3198 *   -0.1315 NS -0.1518 NS 0.1952 NS   -0.2591 NS 0.1094 NS   -0.4662 **
PH  0.3087 *    0.2188 NS   0.3819 **   -0.0202 NS 0.0301 NS   -0.0322 NS 0.1881 NS
EH  0.4782 **   0.1509 NS   0.3024 *    0.1918 NS   -0.0842 NS -0.3159 *   0.3081 *
EPP -0.0261 NS 0.0888 NS   -0.1189 NS -0.1771 NS 0.2856 NS   0.0775 NS   0.3777 *
EL  0.4345 **   -0.1946 NS 0.5402 **   -0.1199 NS -0.3229 *   0.0623 NS   0.59 **
KPR 0.4906 **   0.1838 NS   0.9239 **   0.0063 NS   -0.2117 NS 0.1905 NS   0.5012 **
EG  1 **        0.4554 **   0.5784 **   -0.2214 NS -0.0075 NS 0.1467 NS   0.7528 **
KRP 0.4554 **   1 **        0.5328 **   0.1371 NS   0.1946 NS   0.0069 NS   0.2107 NS
NKE 0.5784 **   0.5328 **   1 **        0.0627 NS   -0.1154 NS 0.1532 NS   0.4994 **
Pro -0.2214 NS 0.1371 NS   0.0627 NS   1 **        -0.0011 NS -0.2548 NS -0.2573 NS
TSS -0.0075 NS 0.1946 NS   -0.1154 NS -0.0011 NS 1 **        -0.1266 NS 0.031 NS
Car 0.1467 NS   0.0069 NS   0.1532 NS   -0.2548 NS -0.1266 NS 1 **        0.1999 NS
GCY 0.7528 **   0.2107 NS   0.4994 **   -0.2573 NS 0.031 NS    0.1999 NS   1 **

$Note1
[1] "The sig of genotypic correlation was tested using t test (two-tail). The degree of free
dom used is number of genotypes - 2"

$Note2
[1] "If NaNs are produced checkout for negative genotypic variance for one or more traits"
```

### *Phenotypic Correlation with 5% and 1% Level of Significance*

Code: p_corr=pheno.corr(data[3:16],data$GEN,data$REP)

**In script editor**

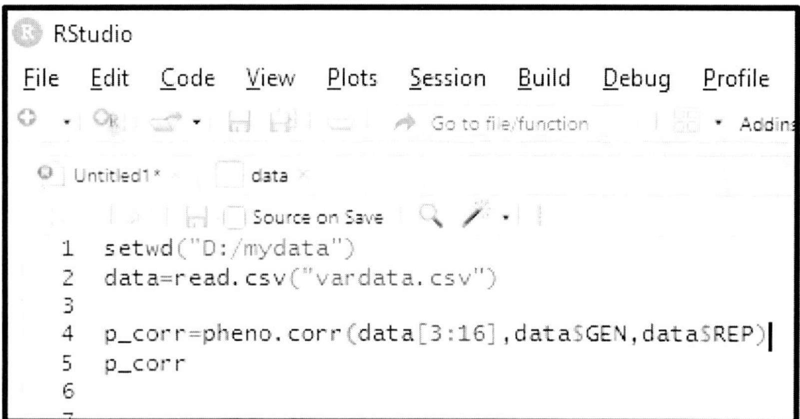


The code for phenotypic correlation analysis is similar to that of genotypic correlation, with the primary difference being the function name, which in this case is "pheno.corr." This function analyzes the same dataset to yield the values of phenotypic correlation.

The results in the console upon running this code.

```
Console  Terminal  Jobs
R 4.0.2 · D:/mydata/
[1] "If NaNs are produced checkout for negative genotypic variance for one or more traits"

> p_corr=pheno.corr(data[3:16],data$GEN,data$REP)
> p_corr
$PhenotypicCorrelation
    TA          SI          PH          EH          EPP         EL          KPR
TA  1 **        0.9604 **   -0.1407 NS  -0.1693 NS  -0.3687 **  -0.2043 NS  -0.1538 NS
SI  0.9604 **   1 **        -0.1978 NS  -0.1985 NS  -0.3853 **  -0.1639 NS  -0.1287 NS
PH  -0.1407 NS  -0.1978 NS  1 **        0.658 **    -0.1631 NS  0.1794 NS   0.3404 **
EH  -0.1693 NS  -0.1985 NS  0.658 **    1 **        -0.2438 *   0.1926 NS   0.2941 **
EPP -0.3687 **  -0.3853 **  -0.1631 NS  -0.2438 *   1 **        -0.0738 NS  -0.1459 NS
EL  -0.2043 NS  -0.1639 NS  0.1794 NS   0.1926 NS   -0.0738 NS  1 **        0.631 **
KPR -0.1538 NS  -0.1287 NS  0.3404 **   0.2941 **   -0.1459 NS  0.631 **    1 **
EG  -0.2314 *   -0.2287 *   0.2855 **   0.3701 **   -0.1172 NS  0.4775 **   0.3788 **
KRP -0.0805 NS  -0.1036 NS  0.214 *     0.1175 NS   0.0898 NS   -0.0851 NS  0.1556 NS
NKE -0.1685 NS  -0.1606 NS  0.35 **     0.276 **    -0.0676 NS  0.506 **    0.9044 **
Pro 0.1488 NS   0.1699 NS   -2e-04 NS   0.1733 NS   -0.0484 NS  -0.0859 NS  0.0067 NS
TSS -0.2551 *   -0.2317 *   0.0094 NS   -0.0855 NS  0.2377 *    -0.2728 **  -0.1974 NS
Car 0.1001 NS   0.1058 NS   -0.029 NS   -0.2952 **  0.0484 NS   0.0528 NS   0.1697 NS
GCY -0.4366 **  -0.3833 **  0.2512 *    0.2745 **   0.2799 **   0.4731 **   0.4586 **
    EG          KRP         NKE         Pro         TSS         Car         GCY
TA  -0.2314 *   -0.0805 NS  -0.1685 NS  0.1488 NS   -0.2551 *   0.1001 NS   -0.4366 **
SI  -0.2287 *   -0.1036 NS  -0.1606 NS  0.1699 NS   -0.2317 *   0.1058 NS   -0.3833 **
PH  0.2855 **   0.214 *     0.35 **     -2e-04 NS   0.0094 NS   -0.029 NS   0.2512 *
EH  0.3701 **   0.1175 NS   0.276 **    0.1733 NS   -0.0855 NS  -0.2952 **  0.2745 **
EPP -0.1172 NS  0.0898 NS   -0.0676 NS  -0.0484 NS  0.2377 *    0.0484 NS   0.2799 **
EL  0.4775 **   -0.0851 NS  0.506 **    -0.0859 NS  -0.2728 **  0.0528 NS   0.4731 **
KPR 0.3788 **   0.1556 NS   0.9044 **   0.0067 NS   -0.1974 NS  0.1697 NS   0.4586 **
EG  1 **        0.3815 **   0.4736 **   -0.1116 NS  -0.0543 NS  0.0888 NS   0.5384 **
KRP 0.3815 **   1 **        0.5494 **   0.0989 NS   0.1728 NS   0.0128 NS   0.2092 *
NKE 0.4736 **   0.5494 **   1 **        0.0514 NS   -0.0987 NS  0.1367 NS   0.4654 **
Pro -0.1116 NS  0.0989 NS   0.0514 NS   1 **        -0.014 NS   -0.2249 *   -0.1818 NS
TSS -0.0543 NS  0.1728 NS   -0.0987 NS  -0.014 NS   1 **        -0.1207 NS  0.0111 NS
Car 0.0888 NS   0.0128 NS   0.1367 NS   -0.2249 *   -0.1207 NS  1 **        0.1586 NS
GCY 0.5384 **   0.2092 *    0.4654 **   -0.1818 NS  0.0111 NS   0.1586 NS   1 **

$Note
[1] "The sig of phenotypic correlation was tested using t test (two-tail). The degree of fre
edom used is (genotypes*replication) - 2"

>
```

## Path Analysis in R

For conducting path analysis, particularly in genotypic and phenotypic path analysis within plant breeding, the dataset requirements are similar to those used in genotypic and phenotypic correlation analysis. This ensures consistency and relevance in examining the relationships between different variables.

### *Code for Genotypic Path Analysis*

Code: g_path=geno.path(data[16],data[3:15],data$GEN,data$REP)

- **g_path:** name of subset where effects of path analysis is saved
- **geno.path:** name of function
- **data[16]:** column address of dependent variable eg. Yield
- **data[3:15]:** column address of independent variables
- **data$GEN:** genotypic or treatment vector
- **data$REP:** Replication vector

**Code in script editor**

RStudio

File Edit Code View Plots Session Build Debug Profile Tools Help

Go to file/function Addins

Untitled1* data

Source on Save

```
setwd("D:/mydata")
data=read.csv("vardata.csv")

g_path=geno.path(data[16],data[3:15],data$GEN,data$REP)
g_path
```

## Result in console

```
Console  Terminal  Jobs
R 4.0.2 · D:/mydata/
> g_path
$Effects
          TA       SI       PH       EH      EPP       EL      KPR       EG      KRP
TA  -1.75363  1.94260 -0.00367 -0.08012 -0.32655 -0.15570  0.12023 -0.10172  0.01749
SI  -1.72426  1.97569 -0.00557 -0.09358 -0.35363 -0.13877  0.09935 -0.10090  0.02284
PH   0.25463 -0.43586  0.02527  0.31589 -0.15798  0.15281 -0.36746  0.09739 -0.03800
EH   0.30138 -0.39656  0.01712  0.46620 -0.23665  0.13905 -0.29645  0.15088 -0.02621
EPP  0.80143 -0.97779 -0.00559 -0.15440  0.71453 -0.10224  0.18613 -0.00823 -0.01543
EL   0.44311 -0.44491  0.00627  0.10520 -0.11856  0.61621 -0.68098  0.13709  0.03379
KPR  0.22169 -0.20640  0.00976  0.14533 -0.13984  0.44124 -0.95103  0.15478 -0.03192
EG   0.56537 -0.63187  0.00780  0.22295 -0.01863  0.26774 -0.46656  0.31550 -0.07908
KRP  0.17661 -0.25988  0.00553  0.07036  0.06349 -0.11990 -0.17478  0.14367 -0.17366
NKE  0.27044 -0.29985  0.00965  0.14099 -0.08497  0.33290 -0.87865  0.18248 -0.09253
Pro -0.34717  0.38573 -0.00051  0.08940 -0.12655 -0.07391 -0.00594 -0.06986 -0.02381
TSS  0.48623 -0.51196  0.00076 -0.03924  0.20409 -0.19895  0.20131 -0.00236 -0.03380
Car -0.17396  0.21620 -0.00081 -0.14725  0.05541  0.03841 -0.18121  0.04628 -0.00119
         NKE      Pro      TSS      Car
TA  -0.13898 -0.01820 -0.01477  0.01917
SI  -0.13677 -0.01795 -0.01380  0.02115
PH   0.34416  0.00186  0.00160 -0.00623
EH   0.27254 -0.01763 -0.00448 -0.06105
EPP -0.10716  0.01628  0.01521  0.01499
EL   0.48686  0.01103 -0.01719  0.01205
KPR  0.83261 -0.00057 -0.01127  0.03683
EG   0.52123  0.02036 -0.00040  0.02835
KRP  0.48016 -0.01260  0.01036  0.00133
NKE  0.90119 -0.00577 -0.00614  0.02961
Pro  0.05655 -0.09194 -0.00006 -0.04924
TSS -0.10397  0.00010  0.05326 -0.02448
Car  0.13807  0.02342 -0.00674  0.19328

$Residual
      [,1]
[1,] 0.035

>
```

## *Phenotypic Path Analysis*

Phenotypic path analysis also requires the same dataset as used in genotypic and phenotypic correlation analysis.

Code: p_path=pheno.path(data[16],data[3:15],data$GEN,data$REP)

```
RStudio
File  Edit  Code  View  Plots  Session  Build  Debug  Profile  Tools  Help
Go to file/function    Addins
Untitled1*    data
Source on Save
setwd("D:/mydata")
data=read.csv("vardata.csv")

p_path=pheno.path(data[16],data[3:15],data$GEN,data$REP)
p_path
```

The code for phenotypic path analysis is similar to that used for genotypic path analysis, with the main difference being the function name. Use the function "pheno.path" to perform phenotypic path analysis.

After running this code, the results will be displayed in the console.

```
Console   Terminal   Jobs
R 4.0.2 · D:/mydata/
> p_path
$Effects
          TA       SI       PH       EH      EPP       EL      KPR       EG      KRP
TA  -0.75453  0.62460 -0.00008 -0.03549 -0.13699 -0.02874 -0.10530 -0.06848 -0.02297
SI  -0.72465  0.65035 -0.00012 -0.04161 -0.14323 -0.02306 -0.08812 -0.06765 -0.02958
PH   0.10616 -0.12857  0.00060  0.13793 -0.06063  0.02524  0.23306  0.08445  0.06105
EH   0.12774 -0.12909  0.00039  0.20962 -0.09063  0.02710  0.20136  0.10948  0.03352
EPP  0.27804 -0.25058 -0.00010 -0.05111  0.37353 -0.01031 -0.09975 -0.03470  0.02565
EL   0.15415 -0.10659  0.00011  0.04037 -0.02725  0.14069  0.43202  0.14124 -0.02428
KPR  0.11605 -0.08370  0.00020  0.06165 -0.05416  0.08877  0.68466  0.11205  0.04439
EG   0.17467 -0.14873  0.00017  0.07758 -0.04361  0.06718  0.25935  0.29583  0.10878
KRP  0.06074 -0.06744  0.00013  0.02463  0.03342 -0.01197  0.10653  0.11279  0.28529
NKE  0.12714 -0.10445  0.00021  0.05786 -0.02513  0.07119  0.61920  0.14009  0.15674
Pro -0.11227  0.11049  0.00000  0.03633 -0.01773 -0.01208  0.00459 -0.03307  0.02827
TSS  0.19263 -0.15056  0.00001 -0.01792  0.08792 -0.03835 -0.13522 -0.01615  0.04930
Car -0.07553  0.06881 -0.00002 -0.06188  0.01807  0.00743  0.11619  0.02627  0.00365
         NKE      Pro      TSS      Car
TA   0.09869 -0.01975  0.00194  0.01050
SI   0.09407 -0.02256  0.00176  0.01110
PH  -0.20500  0.00003 -0.00007 -0.00304
EH  -0.16166 -0.02301  0.00065 -0.03097
EPP  0.03959  0.00633 -0.00179  0.00510
EL  -0.29637  0.01140  0.00207  0.00554
KPR -0.52972 -0.00089  0.00150  0.01780
EG  -0.27739  0.01484  0.00041  0.00932
KRP -0.32179 -0.01316 -0.00131  0.00134
NKE -0.58571 -0.00682  0.00075  0.01434
Pro -0.03011 -0.13273  0.00011 -0.02359
TSS  0.05781  0.00192 -0.00759 -0.01269
Car -0.08007  0.02986  0.00092  0.10491

$Residual
       [,1]
[1,] 0.3905

> |
```

# 7

# Transformation

This chapter begins by providing an essential overview of different data transformation methods and their importance. It primarily concentrates on exploring three crucial transformation techniques: the square root transformation, log transformation, and arc sine transformation. Additionally, this chapter explores the practical application and analysis of these transformations using the R programming language.

## Data Transformation

Data transformation is often needed in statistical analysis and data processing for several key reasons:

- Assumption of homogeneity of variances: ANOVA requires that the variances within each group being compared are equal. When this assumption is violated, i.e., when variances are significantly different across groups, data transformation can help stabilize these variances.
- Normality assumption: ANOVA assumes that the residuals (the differences between observed and predicted values) are normally distributed. However, real-world data often deviates from this normal distribution. Data transformation, such as log or square root transformations, can help in reshaping the distribution of data.
- Handling skewed data: In situations where the data is highly skewed, the results of ANOVA can be misleading. Transformations can reduce skewness, making the distribution more symmetric and suitable for ANOVA.
- Proportional or percentage data: When dealing with data in the form of proportions or percentages, certain transformations (like the arc sine transformation) are necessary. This is because proportions have a bounded scale (0–1 or 0–100%) and their variances are not constant across the range, which can violate ANOVA assumptions.

## Methods of Data Transformation

It covers three key methods to stabilize variance and normalize data distributions in research:

- Square root transformation: Ideal for counting data like the number of infested plants or insects in traps, this transformation is suitable when variance grows with the mean. It's also used for percentage data ranging from 0–20% or 80–100%, after adjusting percentages above 80% by subtracting from 100. When data contains small values, especially with zeros, adding 0.5 before transformation improves accuracy.
- Arcsine (inverse sine) transformation: Best for proportional data, including percentages derived from counts such as barren tillers. It distinguishes between percentages from counts and other types like protein content. For 0 and 100% data points, adjustments are made using a function of the sample size to avoid distortions. This transformation is advised when percentages don't fall between 30 and 70% and are not suitable for square root transformation.
- Logarithmic transformation: Applied to data where the standard deviation varies with the mean, such as the number of insects per plot. It's appropriate for broad-ranging count data. For small values, a unit is added before taking the logarithm to prevent distortion due to low numbers.

## Data Transformation in R

For performing these transformations in R, the package "doebioresearch" is recommended.

Regarding the data file, the format required is .csv, which allows for straightforward reading into R.The data should be organized with a specific structure for optimal processing.

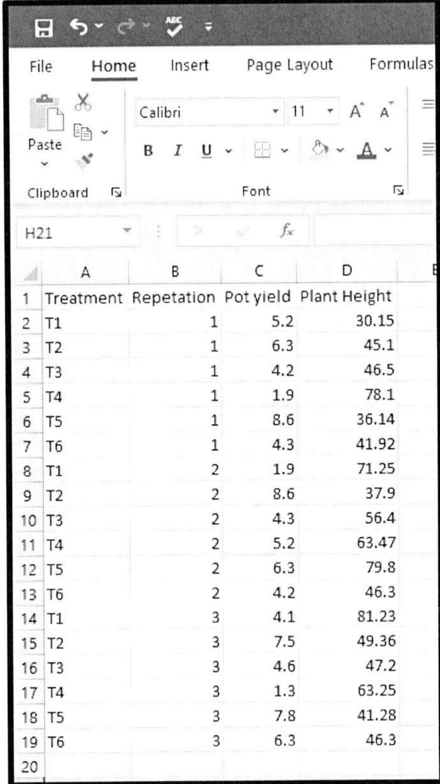

| Treatment | Repetation | Pot yield | Plant Height |
|---|---|---|---|
| T1 | 1 | 5.2 | 30.15 |
| T2 | 1 | 6.3 | 45.1 |
| T3 | 1 | 4.2 | 46.5 |
| T4 | 1 | 1.9 | 78.1 |
| T5 | 1 | 8.6 | 36.14 |
| T6 | 1 | 4.3 | 41.92 |
| T1 | 2 | 1.9 | 71.25 |
| T2 | 2 | 8.6 | 37.9 |
| T3 | 2 | 4.3 | 56.4 |
| T4 | 2 | 5.2 | 63.47 |
| T5 | 2 | 6.3 | 79.8 |
| T6 | 2 | 4.2 | 46.3 |
| T1 | 3 | 4.1 | 81.23 |
| T2 | 3 | 7.5 | 49.36 |
| T3 | 3 | 4.6 | 47.2 |
| T4 | 3 | 1.3 | 63.25 |
| T5 | 3 | 7.8 | 41.28 |
| T6 | 3 | 6.3 | 46.3 |

Make working directory and read this file in RStudio described in earlier chapter.

Now download the package "doebioresearch" and launch it for our analysis.

Take a look at data in the console

```
Console   Terminal   Jobs
R 4.0.2 · D:/mydata/
> data=read.csv("CRD.csv")
> data
  Treatment Repetation Pot.yield Plant.Height
1        T1          1       5.2        30.15
2        T2          1       6.3        45.10
3        T3          1       4.2        46.50
4        T4          1       1.9        78.10
5        T5          1       8.6        36.14
6        T6          1       4.3        41.92
7        T1          2       1.9        71.25
8        T2          2       8.6        37.90
```

### i. Square root transformation

To apply a square root transformation to your data in R, you can use the following code in the script editor. This code performs the transformation on each value of the specified vector. In cases where a value within the vector is 0, an increment of 0.5 is added to each observation to adjust for the transformation:

Code: **transformed_data=sqrttransform(data$Pot.yield)**

If values are zero

Code: **transformed_data = sqrt(data$Pot.yield + 0.5)**

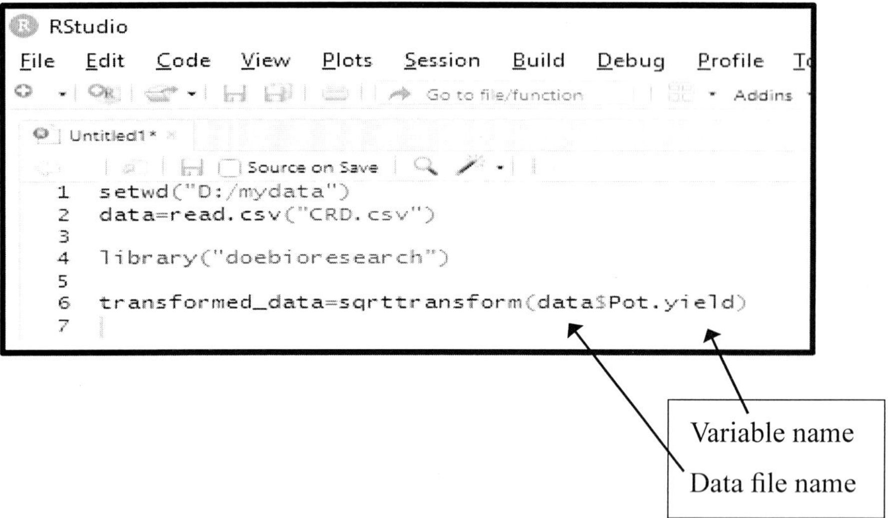


**Result in console**

```
Console  Terminal  Jobs
R 4.0.2 · D:/mydata/
11         T5         2        6.3        79.80
12         T6         2        4.2        46.30
13         T1         3        4.1        81.23
14         T2         3        7.5        49.36
15         T3         3        4.6        47.20
16         T4         3        1.3        63.25
17         T5         3        7.8        41.28
18         T6         3        6.3        46.30
> transformed_data
$Transformed.Data
 [1] 2.280 2.510 2.049 1.378 2.933 2.074 1.378 2.933 2.074 2.280 2.510 2.049 2.025 2.739
[15] 2.145 1.140 2.793 2.510

$Comment
[1] "Zero was not found in the data so no changes were made during tansformation"

>
```

Data of Pot yield is transformed and stored in subset named "transformed_data"

We can get an output of this result using the "sink" code described in Chapter 3.

ii. **Log transformation**

Write the code for log transformation in the script editor

Code: **transformed_data1= logtransform (data$Pot.yield)**

Description of function:

This function applies a base 10 logarithmic transformation to each element in the vector. To accommodate any zero values within the vector, 1 is added to all observations prior to transformation. This log transformation is particularly useful when the data's variance increases with the square of the mean, or when the effects under study are inherently multiplicative.

**Write code in script editor**

```
RStudio
File Edit Code View Plots Session Build Debug Profile Tools Help
Go to file/function  Addins
Untitled1*
Source on Save
setwd("D:/mydata")
data=read.csv("CRD.csv")

library("doebioresearch")

transformed_data1=logtransform(data$Pot.yield)
transformed_data1
```

**Result in console**

```
Console Terminal Jobs
R 4.0.2 · D:/mydata/
> transformed_data1=logtransform(data$Pot.yield)
> transformed_data1
$Ratio
[1] 6.615385

$LogTransformedVector
 [1] 0.7160 0.7993 0.6232 0.2788 0.9345 0.6335 0.2788 0.9345 0.6335 0.7160 0.7993 0.6232
[13] 0.6128 0.8751 0.6628 0.1139 0.8921 0.7993

$Comment
[1] "No observation was found to be zero"

>
```

Description of output

- **Ratio:** The maximum value divided by the minimum value in the dataset.
- **LogTransformedVector:** The series of data points after the log transformation has been applied.
- **Comment:** A note indicating whether the original data contained any zero values.

iii. **Arc sin transformation**

Write the code for log transformation in the script editor

Code: **transformed_data2 = logtransform (data$Pot.yield)**

Description of function

This function performs several operations on a data vector: it first divides each value by 100, then calculates the square root, followed by the sine inverse (arcsine). For any values in the vector that are 0 or 100, they are substituted with $1/(4n)$ or $100 - (1/(4n))$, respectively, where "$n$" is the sample size.

**Write code in the script editor**

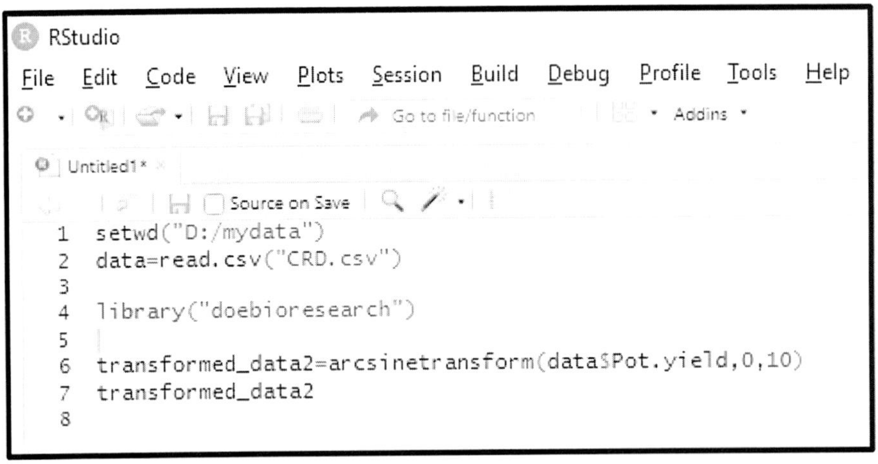

RStudio

File Edit Code View Plots Session Build Debug Profile Tools Help

Go to file/function Addins

Untitled1*

Source on Save

```
setwd("D:/mydata")
data=read.csv("CRD.csv")

library("doebioresearch")

transformed_data2=arcsinetransform(data$Pot.yield,0,10)
transformed_data2

```

- **data$pot.yield:** The vector from the dataset that is subject to transformation.
- **type:** Indicator set to 0 for data expressed as a percentage, and 1 for data expressed as a proportion.
- ***n*:** The sample size.

**Result in console**

```
Console   Terminal   Jobs
R 4.0.2 · D:/mydata/
> transformed_data2=arcsinetransform(data$Pot.yield,0,10)
> transformed_data2
 [1] 13.181417 14.536577 11.825929  7.922904 17.053083 11.967945  7.922904 17.053083
 [9] 11.967945 13.181417 14.536577 11.825929 11.682287 15.894165 12.384793  6.546962
[17] 16.217524 14.536577
>
```

Save this result in a text file using the "sink" function.

# 8

# Completely Randomized Design (CRD)

The chapter on completely randomized design (CRD) presents an overview of its theoretical principles, structure, and randomization methods. It includes the mathematical modeling of CRD and the analysis of ANOVA tables. Additionally, the chapter facilitates an understanding of how to analyze CRD and factorial CRD data using R, offering valuable practical insights for research applications.

## Completely Randomized Design

Completely Randomized Design is a type of experimental design commonly used in statistical analysis and research. It is applied under specific conditions:

- Homogeneous units: CRD is suitable when the experimental units are consistent and uniform across the study area.
- Adherence to experimental design principles: This design conforms to two principal aspects of experimental design i.e. Randomization and Replication.
- Flexibility in replication: It allows for both equal and unequal replication of treatments within the experimental setup.
- Suitable for controlled environments: CRD is particularly effective for experiments conducted in greenhouses or laboratories, where conditions can be tightly regulated.

## Layout

In CRD, the "layout" pertains to how treatments are arranged within the experimental space. For instance, if there are three treatments—A, B, and C—to be replicated eight times, the total experimental space would be divided into 24 equivalent plots. These treatments are then randomly distributed across all plots.

## Layout of three treatments and eight replications

| | | |
|---|---|---|
| B 1 | C 9 | B 17 |
| A 2 | B 10 | A 18 |
| A 3 | C 11 | C 19 |
| A 4 | C 12 | B 20 |
| C 5 | A 13 | A 21 |
| B 6 | A 14 | B 22 |
| A 7 | C 15 | B 23 |
| B 8 | C 16 | C 24 |

## Randomization of Treatments

In the randomization process for treatments within a Completely Randomized Design (CRD) involving 24 units, a two-digit random number table is used. From this table, a sequence of 24 random numbers is selected, omitting any number above 24. For example, consider the random numbers obtained are 4, 18, 2, 21, 14, 3, 7, 13, 22, 1, 6, 10, 17, 23, 20, 8, 15, 11, 24, 5, 9, 12, 16, and 19.

Subsequently, each plot in the experimental area is assigned a sequential number. The treatment allocation is as follows: Treatment A is given to plots numbered 4, 18, 2, 21, 2, 14, 3, 7, and 13; Treatment B is assigned to plots numbered 22, 1, 6, 10, 17, 23, 20, and 8; and the remaining plots receive Treatment C.

An alternative method for randomization is the chit-box approach. In this method, 24 chits, each bearing a number from 1 to 24, are prepared. Chits are drawn randomly to assign plots to Treatment A, followed by Treatments B and C, until all treatments are assigned. This method ensures an unbiased and random distribution of treatments across the experimental plots (Table 1).

Statistical model for CRD: $Y_{ij} = \mu + \tau_i + \varepsilon_{ij}$,

where, $Y_{ij}$ = response of yield from the *j*th unit receiving the *i*th treatment; $\mu$ = general mean;

$\tau_i$ = effect of *i*th treatment; $\varepsilon_{ij}$ = uncontrolled variation associated with *j*th unit receiving *i*th treatment.

**Table 1:** Analysis of variance.

| Source of variation | *DF* | Sum of squares | M.S. | Cal. F |
|---|---|---|---|---|
| Treatment | $(t-1)$ | TSS | TMS = TSS/DF | TMS/EMS |
| Error | $t(r-1)$ | ESS | EMS = ESS/DF | |
| Total | $(rt-1)$ | Total SS | | |

$$SEm = \sqrt{MS_E / r}.$$

CD at 5% if treatment $F$ is significant $= SEm \times \sqrt{2} \times t_{0.05,ne}$

$$CV\% = \frac{\sqrt{MS_E}}{\bar{\bar{Y}}} \times 100.$$

## Factorial Experiment

In simpler experiments, researchers often test variations in a single factor, such as different types of manures, insecticides, or cultural treatments. For these tests, designs like randomized blocks, completely randomized designs, or latin square designs are suitable. However, in practical research, there's often a need to examine the effects of multiple factors simultaneously. For instance, combining irrigation with varying nitrogen doses, or applying insecticides at different times.

In basic experiments, a researcher might first adjust irrigation levels to find the optimum, then do the same for nitrogen doses. However, this approach doesn't consider that these factors might interact and that their combined effects can't be understood by studying them separately. The most effective method for investigating the combined effect of two or more factors is to test all possible combinations of their levels in a single experiment. This method is known as the "Factorial Concept."

Key terminologies:

- Factor: A factor is a set of related treatments. For example, 'nitrogen' is a factor when applying different doses to a crop.
- Levels of a factor: These are the different states or components of a factor. For instance, the various doses of nitrogen represent different levels of the nitrogen factor.

## Factorial CRD

The concept of factorial experiments in completely randomized design (FCRD) extends beyond simple experiments that test a single factor at various levels, such as a varietal trial or a nitrogen level trial. Instead, FCRD involves investigating multiple factors simultaneously within a single experiment, offering a more comprehensive analysis. This approach can be applied not just in CRD, but also in randomized block design (RBD) and Latin square design.

Statistical model for FCRD: $\boldsymbol{Y_{ijk} = \mu + A_j + B_k + AB_{jk} + \varepsilon_{ijk}}$

where, $i = 1, 2, \ldots, r$ and $j = 1, 2, \ldots, v$ and $k = 1, 2, \ldots, n$

Analysis of variance for FCRD

| **Source** | **D.F.** | **SS** | **M. S.** | **Cal *F*** |
|---|---|---|---|---|
| Factor A | $(a - 1)$ | $SS_A$ | $SS_A/DF = MS(A)$ | $MS_A/MSE$ |
| Factor B | $(b - 1)$ | $SS_B$ | $SS_B/ DF = MS_B$ | $MS_B/MSE$ |
| A x B | $(a - 1)(b - 1)$ | $SS_{A \times B}$ | $SS_{AxB}/DF = MS_{AXB}$ | $MS_{AXB}/MSE$ |
| Error | $ab(r - 1)$ | SSE | SSE/DF = MSE | |
| Total | $(abr - 1)$ | TSS | | |

## Analysis Using R

*Title:* "doebioresearch: Comprehensive Analysis of Experimental Designs in Biology."

*Description*: The "doebioresearch" package is tailored for the analysis of prevalent experimental designs in biological research. It encompasses a broad range of designs, including completely randomized design (CRD), randomized complete block design (RCBD), Factorial CRD, Factorial RCBD, split-plot design, strip plot design, and Latin square design.

Key features:

- ✓ Package includes tools for performing analysis of variance (ANOVA), determining the coefficient of determination, conducting normality tests on residuals, and calculating the standard error of the mean and standard error of the difference.
- ✓ It offers capabilities for multiple comparison testing of means.
- ✓ The package is equipped with functions for data transformation and conversion of yield data, enhancing its utility in research.

✓ It also includes several datasets to provide practical examples for users.

To utilize the package, install and launch it in RStudio

## CRD: Completely Randomized Design in R

### *Required Data Format*

The data should be stored in a .csv (comma-separated values) file format, arranged in the following order:

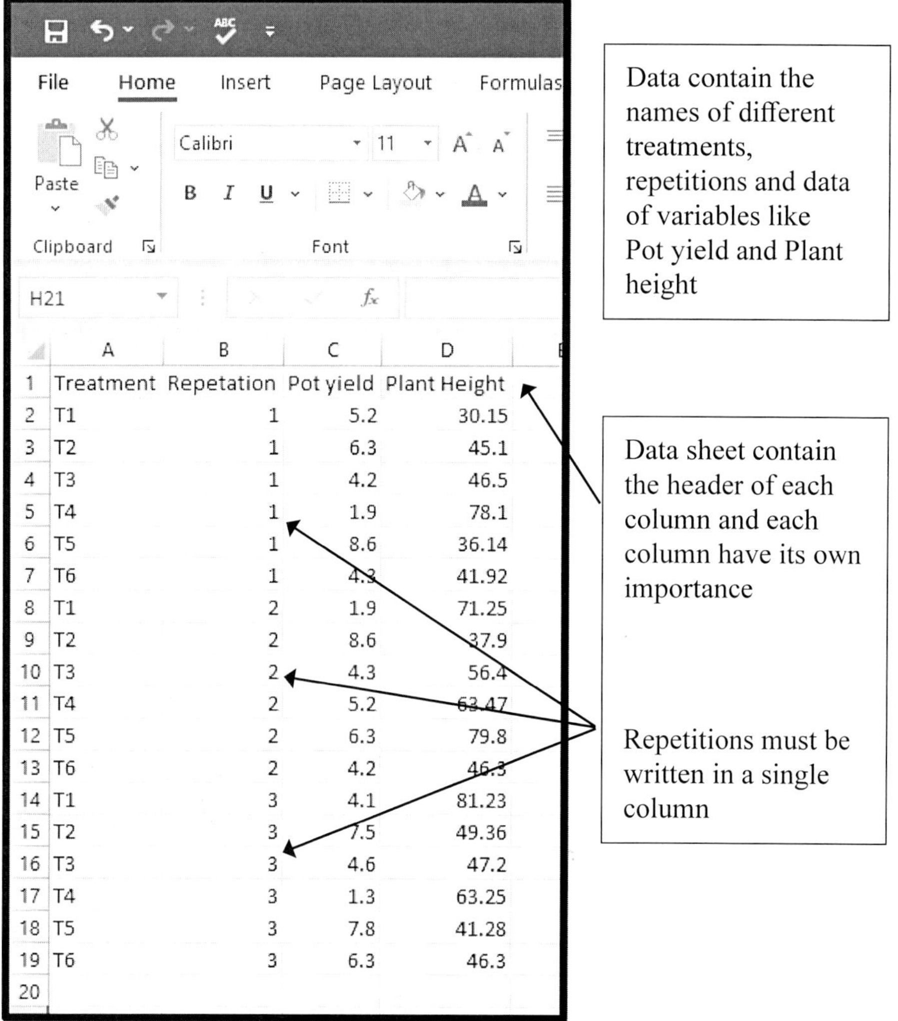


| | A | B | C | D |
|---|---|---|---|---|
| 1 | Treatment | Repetation | Pot yield | Plant Height |
| 2 | T1 | 1 | 5.2 | 30.15 |
| 3 | T2 | 1 | 6.3 | 45.1 |
| 4 | T3 | 1 | 4.2 | 46.5 |
| 5 | T4 | 1 | 1.9 | 78.1 |
| 6 | T5 | 1 | 8.6 | 36.14 |
| 7 | T6 | 1 | 4.3 | 41.92 |
| 8 | T1 | 2 | 1.9 | 71.25 |
| 9 | T2 | 2 | 8.6 | 37.9 |
| 10 | T3 | 2 | 4.3 | 56.4 |
| 11 | T4 | 2 | 5.2 | 63.47 |
| 12 | T5 | 2 | 6.3 | 79.8 |
| 13 | T6 | 2 | 4.2 | 46.3 |
| 14 | T1 | 3 | 4.1 | 81.23 |
| 15 | T2 | 3 | 7.5 | 49.36 |
| 16 | T3 | 3 | 4.6 | 47.2 |
| 17 | T4 | 3 | 1.3 | 63.25 |
| 18 | T5 | 3 | 7.8 | 41.28 |
| 19 | T6 | 3 | 6.3 | 46.3 |
| 20 | | | | |

## Code for CRD analysis

CRD_ANA=crd(data[3:4],data$Treatment,0)

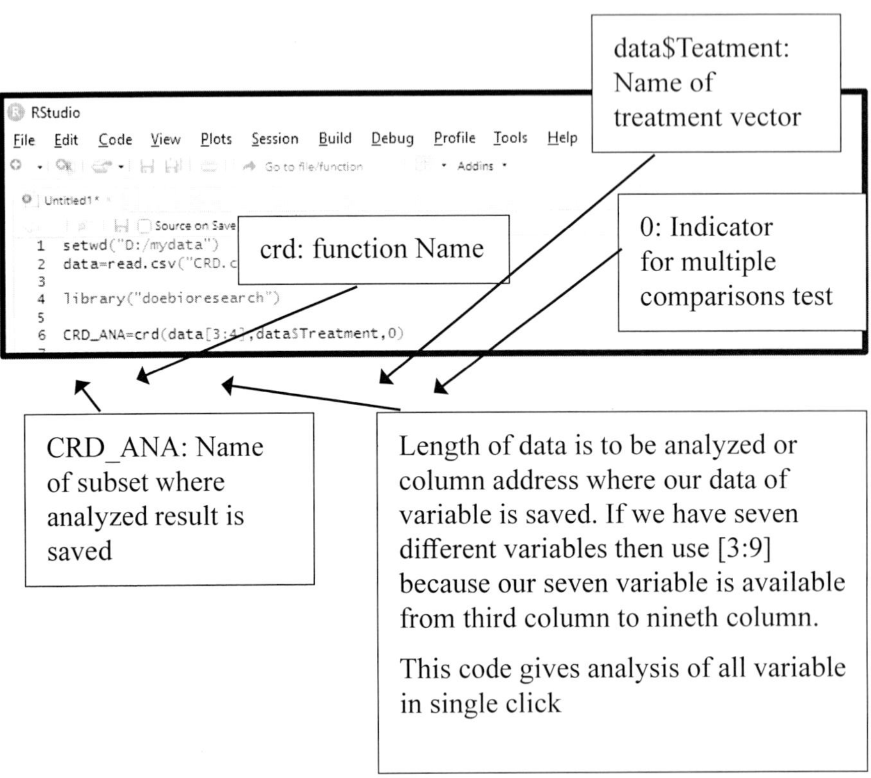


### Description of "crd" function

- **data:** dependent variables
- **trt.vector:** vector containing treatments
- **MultipleComparisonTest:** 0 for no test, 1 for the LSD test, 2 for the Duncan test, and 3 for the HSD test

## Output in console

```
Console   Terminal   Jobs
R 4.0.2 · D:/mydata/

> CRD_ANA=crd(data[3:4],data$Treatment,0)
> CRD_ANA
$Pot.yield
$Pot.yield[[1]]
$Pot.yield[[1]][[1]]
Analysis of Variance Table

Response: data2
          Df Sum Sq Mean Sq F value   Pr(>F)
trt        5 58.191 11.6382  6.1433 0.004771 **
Residuals 12 22.733  1.8944
---
Signif. codes:  0 '***' 0.001 '**' 0.01 '*' 0.05 '.' 0.1 ' ' 1

$Pot.yield[[1]][[2]]
[1] "The treatment means of one or more treatments are not same, so go for multiple comparis
```

Instructions for saving output to a .txt file using the "sink" function:

To capture and download the output of your R code, including various statistical values, into a .txt file, follow these steps:

```
7
8  sink("CRD.txt")
9  print(CRD_ANA)
10 sink()
11
```

Here,

- **Line no 8:** It makes the text file in our working directory with the name "CRD.txt"
- **Line no 9:** It prints the result available in subset "CRD_ANA" from the console
- **Line no 10:** It helps to complete the writing of the result in the text file CRD.txt

## Result in a Text file

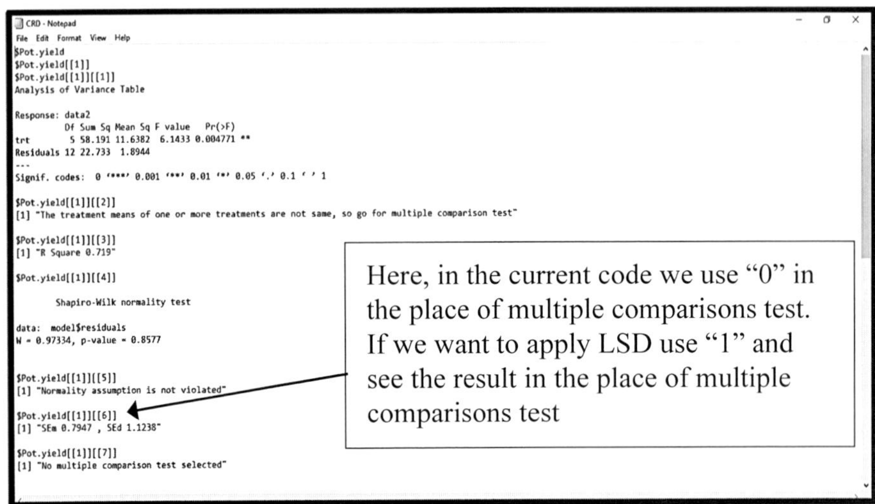

```
$Pot.yield
$Pot.yield[[1]]
$Pot.yield[[1]][[1]]
Analysis of Variance Table

Response: data2
          Df Sum Sq Mean Sq F value   Pr(>F)
trt        5 58.191 11.6382  6.1433 0.004771 **
Residuals 12 22.733  1.8944
---
Signif. codes:  0 '***' 0.001 '**' 0.01 '*' 0.05 '.' 0.1 ' ' 1

$Pot.yield[[1]][[2]]
[1] "The treatment means of one or more treatments are not same, so go for multiple comparison test"

$Pot.yield[[1]][[3]]
[1] "R Square 0.719"

$Pot.yield[[1]][[4]]

        Shapiro-Wilk normality test

data:  model$residuals
W = 0.97334, p-value = 0.8577


$Pot.yield[[1]][[5]]
[1] "Normality assumption is not violated"

$Pot.yield[[1]][[6]]
[1] "SEm 0.7947 , SEd 1.1238"

$Pot.yield[[1]][[7]]
[1] "No multiple comparison test selected"
```

Here, in the current code we use "0" in the place of multiple comparisons test. If we want to apply LSD use "1" and see the result in the place of multiple comparisons test

This file contains the results like ANOVA, interpretation of ANOVA, R-square, normality test result, $SE_m$ and $SE_d$ of all selected variables.

**After applying multiple comparison test**

Code:

```
CRD_ANA=crd(data[3:4],data$Treatment,1)
```

Output in console (Here, we give display for only multiple comparison tests).

```
$Pot.yield[[1]][[7]][[2]]
      data2 groups
T5 7.566667      a
T2 7.466667      a
T6 4.933333      b
T3 4.366667      b
T1 3.733333      b
T4 2.800000      b
```

Here, LSD test is applied and it gives two group indicated by "a" and "b"

## FCRD: Factorial Completely Randomized Design in R

### *Factorial CRD (two and three factors)*

### Required data format

The data should be stored in a .csv (comma-separated values) file format, arranged in the following order:

| | A | B | C | D | E | F |
|---|---|---|---|---|---|---|
| 1 | Replication | Nitrogen | Phosphorus | Potassium | Yield | Plant_Height |
| 2 | 1 | n0 | p0 | k0 | 107 | 10 |
| 3 | 1 | n0 | p0 | k1 | 120 | 12 |
| 4 | 1 | n0 | p1 | k0 | 117 | 11 |
| 5 | 1 | n0 | p1 | k1 | 112 | 16 |
| 6 | 1 | n1 | p0 | k0 | 112 | 10 |
| 7 | 1 | n1 | p0 | k1 | 132 | 15 |
| 8 | 1 | n1 | p1 | k0 | 120 | 10 |
| 9 | 1 | n1 | p1 | k1 | 115 | 12 |
| 10 | 2 | n0 | p0 | k0 | 110 | 8 |
| 11 | 2 | n0 | p0 | k1 | 125 | 9 |
| 12 | 2 | n0 | p1 | k0 | 123 | 6 |
| 13 | 2 | n0 | p1 | k1 | 129 | 12 |
| 14 | 2 | n1 | p0 | k0 | 124 | 13 |
| 15 | 2 | n1 | p0 | k1 | 113 | 14 |
| 16 | 2 | n1 | p1 | k0 | 128 | 11 |
| 17 | 2 | n1 | p1 | k1 | 118 | 9 |
| 18 | 3 | n0 | p0 | k0 | 123 | 12 |
| 19 | 3 | n0 | p0 | k1 | 150 | 15 |
| 20 | 3 | n0 | p1 | k0 | 120 | 14 |
| 21 | 3 | n0 | p1 | k1 | 124 | 16 |
| 22 | 3 | n1 | p0 | k0 | 165 | 17 |
| 23 | 3 | n1 | p0 | k1 | 117 | 18 |
| 24 | 3 | n1 | p1 | k0 | 124 | 15 |
| 25 | 3 | n1 | p1 | k1 | 116 | 13 |
| 26 | | | | | | |

Code for the FCRD with two factors

**Code:**

```
fcrd_ana=fcrd2fact(data[5],data$Nitrogen, data$Phosphorus,1)
```

**Write this code in the script editor**

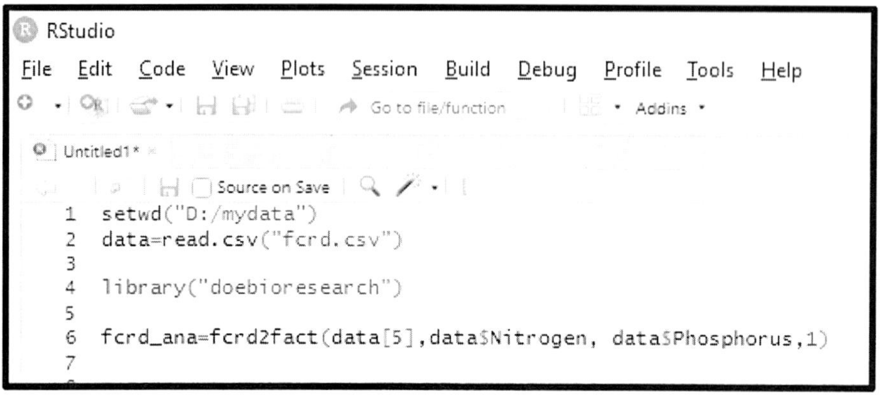

RStudio

File Edit Code View Plots Session Build Debug Profile Tools Help

Go to file/function · Addins

Untitled1*

Source on Save

```
setwd("D:/mydata")
data=read.csv("fcrd.csv")

library("doebioresearch")

fcrd_ana=fcrd2fact(data[5],data$Nitrogen, data$Phosphorus,1)
```

**Description of code**

- ✓ **fcrd_ana:** This is the variable where the results of our analysis are stored.
- ✓ **fcrd2fact:** A function designed for factorial completely randomized design (FCRD) analysis involving two factors.
- ✓ **data[5]:** Refers to the subset of our dataset where our variables are saved, with the fifth column indicating the specific variable. To analyze additional variables, modify this to "data[5:n]," where "*n*" represents the column number of the last variable.
- ✓ **data$Nitrogen:** The first factor in the dataset.
- ✓ **data$Phophorus:** The second factor in the dataset.
- ✓ **1:** for multiple comparison test (LSD), 0 for no comparison, 2 for Duncan test, 3 for HSD test indicator of multiple comparison test (LSD)

## Output in console

```
Console  Terminal  Jobs
R 4.0.2 · D:/mydata/
24          3         n1          p1         k1    116           13
> fcrd_ana=fcrd2fact(data[5],data$Nitrogen, data$Phosphorus,1)
> fcrd_ana
$Yield
$Yield[[1]]
Analysis of Variance Table

Response: dependent.var
               Df Sum Sq Mean Sq F value Pr(>F)
fact.A          1   24.0  24.000  0.1386 0.7136
fact.B          1  112.7 112.667  0.6505 0.4294
fact.A:fact.B   1   42.7  42.667  0.2463 0.6251
Residuals      20 3464.0 173.200

$Yield[[2]]
[1] "R Square 0.049"

$Yield[[3]]
[1] "SEm of A: 3.799 , SEd of A: 5.373 , SEm of B: 3.799 , SEd of B: 5.373 , SEm of AB: 5.37
3 , SEd of AB 7.598"

$Yield[[4]]

        Shapiro-Wilk normality test

data:  model$residuals
W = 0.85265, p-value = 0.002449


$Yield[[5]]
[1] "Normality assumption is violated"

$Yield[[6]]
[1] "All the first factor level means are same so dont go for any multiple comparison test"

$Yield[[7]]
$Yield[[7]][[1]]
  MSerror Df     Mean       CV  t.value     LSD
    173.2 20 122.6667 10.72871 2.085963 11.2074

$Yield[[7]][[2]]
```

## Complete output in a text file

$Yield

$Yield[[1]]

Analysis of Variance Table

Response: dependent.var

| | Df | Sum Sq | Mean Sq | F value | Pr(>F) |
|---|---|---|---|---|---|
| fact.A | 1 | 24.0 | 24.000 | 0.1386 | 0.7136 |
| fact.B | 1 | 112.7 | 112.667 | 0.6505 | 0.4294 |

```
fact.A:fact.B  1   42.7  42.667  0.2463 0.6251
Residuals     20 3464.0 173.200

$Yield[[2]]
[1] "R Square 0.049"

$Yield[[3]]
[1] "SEm of A: 3.799 , SEd of A: 5.373 , SEm of B: 3.799 , SEd of B: 5.373 , SEm of AB: 5.373
, SEd of AB 7.598"

$Yield[[4]]

Shapiro-Wilk normality test

data:  model$residuals
W = 0.85265, p-value = 0.002449


$Yield[[5]]
[1] "Normality assumption is violated"

$Yield[[6]]
[1] "All the first factor level means are same so dont go for any multiple comparison test"

$Yield[[7]]
$Yield[[7]][[1]]
 MSerror Df    Mean       CV  t.value     LSD
   173.2 20 122.6667 10.72871 2.085963 11.2074

$Yield[[7]][[2]]
  dependent.var groups
n1      123.6667      a
n0      121.6667      a
```

```
$Yield[[8]]
[1] "All the second factor level means are same so dont go for any multiple comparison test"


$Yield[[9]]
$Yield[[9]][[1]]
 MSerror Df     Mean       CV  t.value     LSD
   173.2 20 122.6667 10.72871 2.085963 11.2074


$Yield[[9]][[2]]
  dependent.var groups
p0     124.8333      a
p1     120.5000      a



$Yield[[10]]
[1] "The means of levels of interaction between two factors are same so dont go for any multiple comparison test"


$Yield[[11]]
$Yield[[11]][[1]]
 MSerror Df     Mean       CV  t.value      LSD
   173.2 20 122.6667 10.72871 2.085963 15.84966


$Yield[[11]][[2]]
      dependent.var groups
n1:p0      127.1667      a
n0:p0      122.5000      a
n0:p1      120.8333      a
n1:p1      120.1667      a
```

# 9

# Randomized Block Design (RBD)

This chapter provides a comprehensive overview of the randomized block design (RBD), detailing its theoretical concept, layout, and a detailed segment on pooled ANOVA concepts. It explains the mathematical models for RBD and how to interpret the results from ANOVA tables in this context. The chapter also offers step-by-step guidance for analyzing RBD, Factorial RBD, and pooled RBD ANOVA using R.

## Randomized Block Design

The RBD is a widely used experimental design in statistics and research, particularly when dealing with certain specific conditions:

- Homogeneity within blocks: RBD is ideal when there is inherent variability in the experimental units, but this variability can be grouped into homogenous blocks. Each block contains a complete set of treatments, and the blocks themselves account for the variability.
- Principles of experimental design: RBD adheres to the core principles of experimental design, namely randomization, replication, and local control. Randomization is used within blocks to assign treatments, ensuring that the comparison of treatments is fair.
- Replication across blocks: In RBD, each treatment is replicated across different blocks. This replication enhances the reliability and robustness of the experimental results.
- Managing uncontrolled variability: RBD is particularly effective in field experiments or situations where external variability (e.g., soil type, light conditions) cannot be completely controlled. By grouping similar units into blocks, RBD effectively reduces the impact of this uncontrolled variability on the treatment effects.
- Improvement in experimental accuracy: The use of blocks to organize the experimental units usually leads to a reduction in the error variance compared to CRD. This enhances the accuracy and precision of the experiment.

## Layout of RBD

In RBD, the "layout" pertains to how treatments are arranged within each block. Consider an experiment with five treatments labeled A, B, C, D, and E, each replicated four times. The entire experimental area is divided into four blocks, each designed to be relatively homogeneous. Within each block, there are five plots. Treatments are randomly assigned to the plots in each block, with a new randomization process conducted for each block. An example of such a layout is explained in Table 1.

**Table 1:** The layout of five treatments and four replications.

| Block I | Block II | Block III | Block IV |
|---|---|---|---|
| B | A | C | D |
| A | C | D | E |
| C | D | E | B |
| D | E | B | A |
| E | B | A | C |

## Randomization of Treatments

To randomize the treatments, Tippet's random number tables can be utilized. By choosing one-digit numbers as they appear in the table and excluding zero and numbers greater than 5, a sequence of random numbers from 1 to 5, such as 2, 1, 3, 4, and 5, can be obtained. For the first block, this sequence would mean allocating Treatment A to the second plot, B to the first plot, C to the third, D to the fourth, and E to the fifth plot. This randomization process is then independently repeated for each of the remaining blocks, ensuring that treatments are assigned to plots in a random but structured manner (Table 2).

Statistical model for RBD: $y_{ij} = \mu + \alpha_i + \beta_j + \varepsilon_{ij}$,

where, $Y_{ij}$ = response of yield from the $j$th unit receiving the $i$th treatment, $\mu$ = general mean, $\alpha_i$ = effect of $i$th treatment, $\beta_j$ = effect due to $j$th block/ replication, $\varepsilon_{ij}$= uncontrolled variation associated with $j$th unit receiving $i$th treatments.

**Table 2:** Analysis of variance.

| Source of variation (S.V.) | Degrees of freedom (*d.f.*) | Sum of square (S.S.) | Mean square (M.S.) | *F*-calculated value | *F*-*t*abulated value |
|---|---|---|---|---|---|
| Replication /block | $r-1$ | R.S.S. | R.M.S. = R.S.S./$r-1$ | R.M.S./ M.S.E. | *F*[r-1, (r-1) (t-1)] at 0.05 level of significance |
| Treatment ($t$) | $t-1$ | T.S.S | T.M.S. = T.S.S./$t-1$ | T.M.S./ M.S.E. | F[t-1, (r-1) (t-1)] at 0.05 level of significance |
| Error | $(r-1)(t-1)$ | E.S.S | M.S.E. = E.S.S. /$(r-1)(t-1)$ | | |
| Total ($n$) | rt-1 | Total S.S. | | | |

$$SEm = \sqrt{MS_E/_r}.$$

CD at 5% if treatment $F$ is significant = $SEm \times \sqrt{2} \times t_{0.05,ne}$.

$$CV\% = \frac{\sqrt{MS_E}}{\overline{Y}} \times 100.$$

## Factorial RBD

Factorial experiments in randomized block design (FRBD) go beyond evaluating a single factor by testing multiple factors concurrently within the same experimental setup. This method enhances the depth of analysis by examining the interactions between different factors while maintaining the advantages of the RBD. In FRBD, each block contains all combinations of the levels of the factors being tested, allowing for a thorough investigation of how these factors interact within different blocks. This comprehensive approach is particularly useful in experiments where block-to-block variability is a concern, as it helps isolate the effects of the factors from potential confounding variables (Table 3).

### *Statistical Model for FCRD*

$Y_{ijkl} = \mu + R_l + A_j + B_k + AB_{jk} + \varepsilon_{ijkl}$, where, $j = 1, 2, \ldots, a$; $k = 1, 2, \ldots, b$ and $l = 1, 2, \ldots, r$.

**Table 3:** Analysis of variance for FCRD.

| Source | *D.F.* | SS | M.S. | Cal *F* |
|---|---|---|---|---|
| Replication | $(r-1)$ | $SS_R$ | $SS_R/DF = MS(R)$ | $MS_R/MSE$ |
| Factor *A* | $(a-1)$ | $SS_A$ | $SS_A/DF = MS(A)$ | $MS_A/MSE$ |
| Factor *B* | $(b-1)$ | $SS_B$ | $SS_B/DF = MS_B$ | $MS_B/MSE$ |
| $A \times B$ | $(a-1)(b-1)$ | $SS_{A\times B}$ | $SS_{A\times B}/ DF = MS_{A\times B}$ | $MS_{A\times B}/MSE$ |
| Error | $ab(r-1)$ | SSE | SSE/DF = MSE | |
| Total | $(abr-1)$ | TSS | | |

## Pooled ANOVA

Pooled ANOVA is a statistical method used to analyze the results of long-term experiments, which are studies conducted over an extended period or across various locations and seasons. These experiments are crucial in agriculture, where crop performance varies significantly due to environmental factors like rainfall and temperature. Repetition across different conditions allows for more reliable conclusions, particularly when testing new crop varieties or techniques.

In setting up these experiments, the same design (such as CRD, RBD, or split-plot) is consistently applied across all locations or seasons. However, treatments are randomized afresh for each specific experiment.

The analysis involves a combined approach:

i. Data from individual environments are analyzed based on the chosen experimental design.

ii. A combined analysis of variance is conducted, compiling all data.

For instance, in an RBD design with "*t*" treatments, "*r*" replications, and "*p*" environments, the ANOVA table includes elements like replication within environments, environments, treatments, environments-treatment interactions, and combined error. The significance of treatment-location interaction is assessed to determine if treatment responses vary with environmental conditions.

| Sources of variation | Degrees of freedom | Sum of Squares | Mean Squares | *F*-ratio |
|---|---|---|---|---|
| Replication within environments | $p(r-1)$ | RSS | RMS | |
| Environments | $p-1$ | LSS | LMS | |
| Treatments | $t-1$ | TrSS | TrMS | TrMS/L*TMS |
| Location × treatments | $(p-1)(t-1)$ | L*TSS | L*TMS | L*TMS/EMS |
| Combined error | $p(r-1)(t-1)$ | ESS | EMS | |
| Total | $rtp-1$ | TSS | | |

Before the combined analysis, it is essential to verify the homogeneity of the error mean squares (EMS) across experiments, using tests like Bartlett's or Hartley's. If EMS is homogeneous, the analysis proceeds with pooled error sums of squares, forming a two-way table to compute treatment, location, and interaction sums of squares.

## RCBD Using R

### *RCBD: Randomized Complete Block Design*

*Required data format*

The data should be stored in a .csv (comma-separated values) file format, arranged in the following order:

| | A | B | C | D |
|---|---|---|---|---|
| 1 | Treatment | Replication | Yield | Plant Height |
| 2 | T1 | 1 | 5.2 | 30.15 |
| 3 | T2 | 1 | 6.3 | 45.1 |
| 4 | T3 | 1 | 4.2 | 46.5 |
| 5 | T4 | 1 | 1.9 | 78.1 |
| 6 | T5 | 1 | 8.6 | 36.14 |
| 7 | T6 | 1 | 4.3 | 41.92 |
| 8 | T1 | 2 | 1.9 | 71.25 |
| 9 | T2 | 2 | 8.6 | 37.9 |
| 10 | T3 | 2 | 4.3 | 56.4 |
| 11 | T4 | 2 | 5.2 | 63.47 |
| 12 | T5 | 2 | 6.3 | 79.8 |
| 13 | T6 | 2 | 4.2 | 46.3 |
| 14 | T1 | 3 | 4.1 | 81.23 |
| 15 | T2 | 3 | 7.5 | 49.36 |
| 16 | T3 | 3 | 4.6 | 47.2 |
| 17 | T4 | 3 | 1.3 | 63.25 |
| 18 | T5 | 3 | 7.8 | 41.28 |
| 19 | T6 | 3 | 6.3 | 46.3 |
| 20 | | | | |

Code for the RCBD analysis

**RBD_ana=rcbd(data[3:4],data$Treatment, data$Replication,1)**

RStudio

File Edit Code View Plots Session Build Debug Profile Tools Help

Go to file/function Addins

Untitled1*

Source on Save

```
setwd("D:/mydata")
data=read.csv("RBD.csv")

library("doebioresearch")

RBD_ana=rcbd(data[3:4],data$Treatment, data$Replication,1)

```

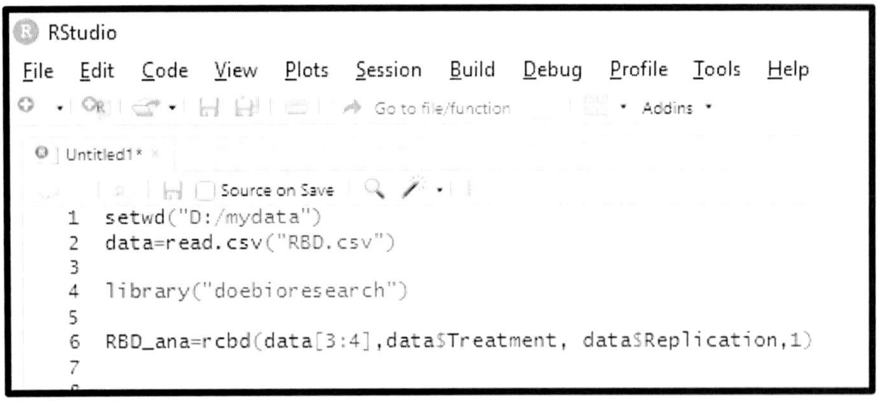

### Description of code

- ✓ **RBD**_ana: This is the subset where the results of the analysis are stored.
- ✓ **data[3:4]**: Represents the subset of the dataset being used.
- ✓ **data$Treatment**: Refers to the column in the dataset that contains the treatment information.
- ✓ **data$Replication**: Denotes the column in the dataset that holds the replication information.
- ✓ **1, 0, 2, 3**: These are indicators for different types of multiple comparison tests. "1" represents the least significant difference (LSD) test, "0" indicates no comparison test

### Console

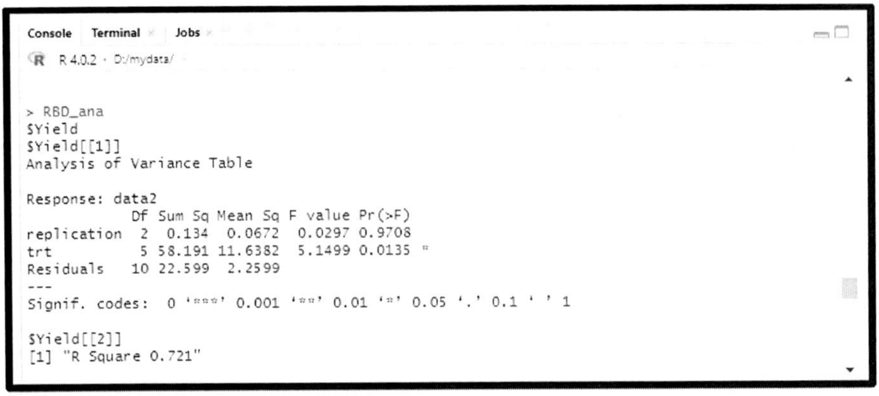

Console Terminal Jobs

R 4.0.2 · D:/mydata/

```
> RBD_ana
$Yield
$Yield[[1]]
Analysis of Variance Table

Response: data2
            Df Sum Sq Mean Sq F value Pr(>F)
replication  2  0.134  0.0672  0.0297 0.9708
trt          5 58.191 11.6382  5.1499 0.0135 *
Residuals   10 22.599  2.2599
---
Signif. codes:  0 '***' 0.001 '**' 0.01 '*' 0.05 '.' 0.1 ' ' 1

$Yield[[2]]
[1] "R Square 0.721"
```

## Complete output in text file

```
$Yield

$Yield[[1]]
Analysis of Variance Table

Response: data2
           Df Sum Sq Mean Sq F value Pr(>F)
replication  2  0.134  0.0672  0.0297 0.9708
trt          5 58.191 11.6382  5.1499 0.0135 *
Residuals   10 22.599  2.2599
---
Signif. codes:  0 '***' 0.001 '**' 0.01 '*' 0.05 '.' 0.1 ' ' 1

$Yield[[2]]
[1] "R Square 0.721"

$Yield[[3]]

        Shapiro-Wilk normality test

data:  model$residuals
W = 0.96518, p-value = 0.7037

$Yield[[4]]
[1] "Normality assumption is not violated"

$Yield[[5]]
[1] "SEm 0.8679 , SEd 1.2274"

$Yield[[6]]
[1] "The treatment means of one or more treatments are not same, so go for multiple comparison test"
```

```
$Yield[[7]]
$Yield[[7]][[1]]
  MSerror Df     Mean       CV  t.value      LSD
 2.259889 10 5.144444 29.22167 2.228139 2.734892

$Yield[[7]][[2]]
     data2 groups
T5 7.566667      a
T2 7.466667      a
T6 4.933333     ab
T3 4.366667      b
T1 3.733333      b
T4 2.800000      b


$Plant.Height
$Plant.Height[[1]]
Analysis of Variance Table

Response: data2
           Df  Sum Sq Mean Sq F value Pr(>F)
replication  2  513.06  256.53  1.0790 0.3764
trt          5 1346.48  269.30  1.1327 0.4037
Residuals   10 2377.54  237.75

$Plant.Height[[2]]
[1] "R Square 0.439"

$Plant.Height[[3]]
```

```
        Shapiro-Wilk normality test


data:  model$residuals
W = 0.97044, p-value = 0.8059



$Plant.Height[[4]]
[1] "Normality assumption is not violated"

$Plant.Height[[5]]
[1] "SEm 8.9023 , SEd 12.5898"

$Plant.Height[[6]]
[1] "All the treatment means are same so dont go for any multiple comparison test"

$Plant.Height[[7]]
$Plant.Height[[7]][[1]]
  MSerror Df  Mean      CV  t.value    LSD
 237.7543 10 53.425 28.86155 2.228139 28.0518

$Plant.Height[[7]][[2]]
     data2 groups
T4 68.27333      a
T1 60.87667      a
T5 52.40667      a
T3 50.03333      a
T6 44.84000      a
T2 44.12000      a
```

## "Variability" package for RCBD

If you are working with the data of genotypes for variability analysis then we need a different package.

When conducting variability analysis with genotype data, a specific R package, named "variability" is required. To use this package, first download and install it, then initiate it within your R environment. For conducting ANOVA and variability analysis effectively, it's important to ensure that your data is formatted correctly to be compatible with the "variability" package.

Name of package: "variability" Download, install, and launch the package.

**Required data format**

The data should be stored in a .csv (comma-separated values) file format, arranged in the following order:

**Read this .csv file in RStudio for analysis**

File Home Insert Page Layout Formul

Paste Clipboard | Calibri 11 B I U Font

F12

| | A | B | C | D | E |
|---|---|---|---|---|---|
| 1 | Treatment | Replicatio | Yield | Plant Height | |
| 2 | Genotype1 | 1 | 5.2 | 30.15 | |
| 3 | Genotype2 | 1 | 6.3 | 45.1 | |
| 4 | Genotype3 | 1 | 4.2 | 46.5 | |
| 5 | Genotype4 | 1 | 1.9 | 78.1 | |
| 6 | Genotype5 | 1 | 8.6 | 36.14 | |
| 7 | Genotype6 | 1 | 4.3 | 41.92 | |
| 8 | Genotype1 | 2 | 1.9 | 71.25 | |
| 9 | Genotype2 | 2 | 8.6 | 37.9 | |
| 10 | Genotype3 | 2 | 4.3 | 56.4 | |
| 11 | Genotype4 | 2 | 5.2 | 63.47 | |
| 12 | Genotype5 | 2 | 6.3 | 79.8 | |
| 13 | Genotype6 | 2 | 4.2 | 46.3 | |
| 14 | Genotype1 | 3 | 4.1 | 81.23 | |
| 15 | Genotype2 | 3 | 7.5 | 49.36 | |
| 16 | Genotype3 | 3 | 4.6 | 47.2 | |
| 17 | Genotype4 | 3 | 1.3 | 63.25 | |
| 18 | Genotype5 | 3 | 7.8 | 41.28 | |
| 19 | Genotype6 | 3 | 6.3 | 46.3 | |
| 20 | | | | | |

**To preview a snippet of the data file in the R console**

```
Console  Terminal  Jobs
R 4.0.2 · D:/mydata/
> data=read.csv("RBD.csv")
> data
   Treatment Replication Yield Plant.Height
1  Genotype1           1   5.2        30.15
2  Genotype2           1   6.3        45.10
3  Genotype3           1   4.2        46.50
4  Genotype4           1   1.9        78.10
5  Genotype5           1   8.6        36.14
6  Genotype6           1   4.3        41.92
7  Genotype1           2   1.9        71.25
8  Genotype2           2   8.6        37.90
9  Genotype3           2   4.3        56.40
10 Genotype4           2   5.2        63.47
11 Genotype5           2   6.3        79.80
12 Genotype6           2   4.2        46.30
13 Genotype1           3   4.1        81.23
14 Genotype2           3   7.5        49.36
15 Genotype3           3   4.6        47.20
16 Genotype4           3   1.3        63.25
17 Genotype5           3   7.8        41.28
18 Genotype6           3   6.3        46.30
```

Code for RCBD analysis with variability analysis of variable

**Code**

```
RBD_var=gen.var(data[3:4],data$Treatment,data$Replication)
```

**Write this code in the script editor for analysis**

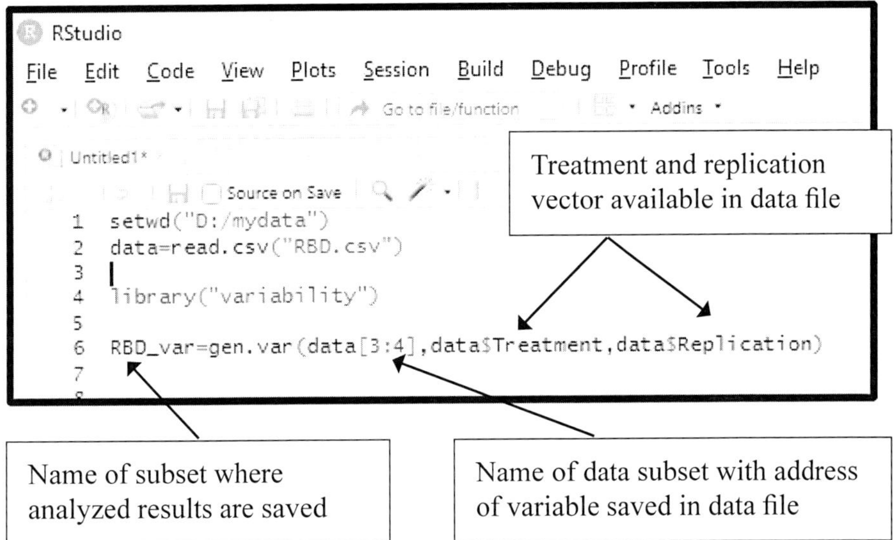

## Console

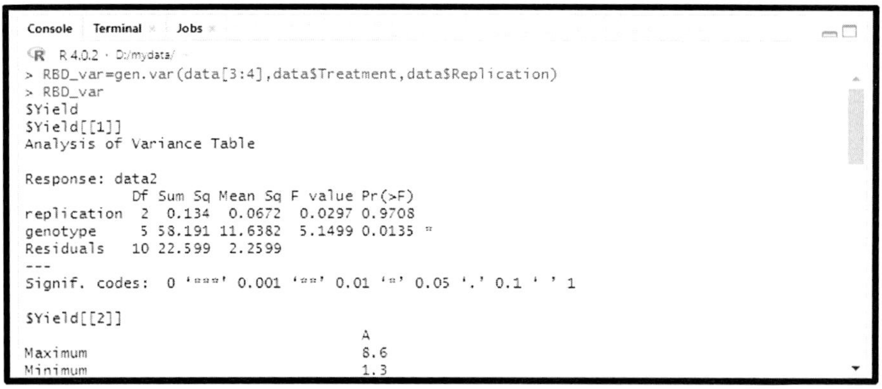

```
Console   Terminal ×   Jobs ×
R 4.0.2 · D:/mydata/
> RBD_var=gen.var(data[3:4],data$Treatment,data$Replication)
> RBD_var
$Yield
$Yield[[1]]
Analysis of Variance Table

Response: data2
            Df Sum Sq Mean Sq F value Pr(>F)
replication  2  0.134  0.0672  0.0297 0.9708
genotype     5 58.191 11.6382  5.1499 0.0135 *
Residuals   10 22.599  2.2599
---
Signif. codes:  0 '***' 0.001 '**' 0.01 '*' 0.05 '.' 0.1 ' ' 1

$Yield[[2]]
                                     A
Maximum                              8.6
Minimum                              1.3
```

## Complete output in text file

```
$Yield

$Yield[[1]]

Analysis of Variance Table


Response: data2

            Df Sum Sq Mean Sq F value Pr(>F)

replication  2  0.134  0.0672  0.0297 0.9708

genotype     5 58.191 11.6382  5.1499 0.0135 *

Residuals   10 22.599  2.2599

---

Signif. codes:  0 '***' 0.001 '**' 0.01 '*' 0.05 '.' 0.1 ' ' 1


$Yield[[2]]

                                        A

Maximum                                 8.6

Minimum                                 1.3

Grand Mean                              5.1444

Standard Error of Mean (SEm)            0.8679

Critical Difference (CD) 5%             2.7349
```

| | |
|---|---|
| Critical Difference (CD) 1% | 3.8901 NS |
| Environmental Variance | 2.2599 |
| Genotypic Variance | 3.1261 |
| Phenotypic Variance | 5.386 |
| Environmental Coefficient of Variance | 29.2217 |
| Genotypic Coefficient of Variance | 34.3687 |
| Phenotypic Coefficient of Variance | 45.1123 |
| Heritability (Broad Sense) | 0.5804 |
| Genetic Advance | 2.7748 |
| Genetic Advance as percentage of mean | 53.9378 |

```
$Plant.Height
$Plant.Height[[1]]
Analysis of Variance Table

Response: data2
         Df  Sum Sq Mean Sq F value Pr(>F)
replication  2  513.06  256.53  1.0790 0.3764
genotype     5 1346.48  269.30  1.1327 0.4037
Residuals   10 2377.54  237.75

$Plant.Height[[2]]
```

| | A |
|---|---|
| Maximum | 81.23 |
| Minimum | 30.15 |
| Grand Mean | 53.425 |
| Standard Error of Mean (SEm) | 8.9023 |
| Critical Difference (CD) 5% | 28.0518 NS |
| Critical Difference (CD) 1% | 39.9005 NS |
| Environmental Variance | 237.7543 |

| | |
|---|---|
| Genotypic Variance | 10.5138 |
| Phenotypic Variance | 248.2681 |
| Environmental Coefficient of Variance | 28.8615 |
| Genotypic Coefficient of Variance | 6.0693 |
| Phenotypic Coefficient of Variance | 29.4928 |
| Heritability (Broad Sense) | 0.0423 |
| Genetic Advance | 1.3746 |
| Genetic Advance as a percentage of mean | 2.573 |

### *Highlights*

The "variability" package in R produces a comprehensive output that includes ANOVA results, the maximum and minimum values across all treatments, the overall grand mean, the standard error of the mean (S.Em.), the critical difference (CD), and various variance measures such as environmental variance, genotypic variance, and phenotypic variance.

It also calculates and provides the environmental, genotypic, and phenotypic coefficients of variance (expressed in percentage), along with broad sense heritability, genetic advance (GA), and genetic advance as a percentage of the mean.

## Factorial RCBD Using R

For further analysis, the "doebioresearch" package in R is necessary. To start using this package, activate it with the command "library ('doebioresearch')." The data format required for this package aligns with that used for factorial completely randomized design (FCRD).

### Code for the FRBD analysis

```
RBD_fact=frbd2fact(data[5],data$Replication,data$Nitrogen,data$Phosphorus,1)
```

### In script editor

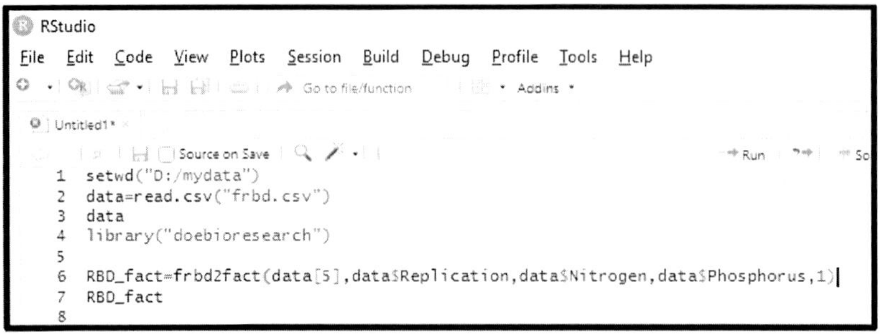
RStudio
File Edit Code View Plots Session Build Debug Profile Tools Help
Go to file/function
Addins
Untitled1
Source on Save
Run
setwd("D:/mydata")
data=read.csv("frbd.csv")
data
library("doebioresearch")
RBD_fact=frbd2fact(data[5],data$Replication,data$Nitrogen,data$Phosphorus,1)
RBD_fact

**Description of code**

- ✓ **RBD_fact**: This subset stores the results of the analysis.
- ✓ **frbd2fact**: Function for FCRD with two-factor.
- ✓ **data[5]:** This subset refers to the specific column in the dataset containing the variable to be analyzed. For analyzing additional variables, expand the range to "data[5:n]," where "*n*" represents the column number of the last variable.
- ✓ **data$Nitrogen:** First factor from data.
- ✓ **data$Phophorus:** Second factor.
- ✓ **1,0,2,3:** These numbers are indicators for various types of multiple comparison tests. "1" stands for the LSD test, "0" means no comparison test is applied, "2" indicates the Duncan test, and "3" is for the Tukey's honestly significant difference (HSD) test.

**Complete output in text file**

$Yield

$Yield[[1]]

Analysis of Variance Table

Response: dependent.var

| | Df | Sum Sq | Mean Sq | F value | Pr(>F) |
|---|---|---|---|---|---|
| replicationvector | 2 | 700.08 | 350.04 | 2.2796 | 0.1311 |
| fact.A | 1 | 24.00 | 24.00 | 0.1563 | 0.6972 |
| fact.B | 1 | 112.67 | 112.67 | 0.7337 | 0.4029 |
| fact.A:fact.B | 1 | 42.67 | 42.67 | 0.2779 | 0.6045 |
| Residuals | 18 | 2763.92 | 153.55 | | |

$Yield[[2]]

[1] "R Square 0.241"

$Yield[[3]]

[1] "SEm of A: 5.059 , SEd of A: 7.154 , SEm of B: 5.059 , SEd of B 7.154 , SEm of AB: 7.154 , SEd of AB: 10.118"

```
$Yield[[4]]

Shapiro-Wilk normality test

data:  model$residuals
W = 0.94271, p-value = 0.1874



$Yield[[5]]
[1] "Normality assumption is not violated"

$Yield[[6]]
[1] "All the first factor level means are same so dont go for any multiple comparison test"

$Yield[[7]]
$Yield[[7]][[1]]
  MSerror Df    Mean       CV  t.value      LSD
 153.5509 18 122.6667 10.10182 2.100922 10.62822

$Yield[[7]][[2]]
   dependent.var groups
n1      123.6667      a
n0      121.6667      a

$Yield[[8]]
[1] "All the second factor level means are same so dont go for any multiple comparison test"

$Yield[[9]]
$Yield[[9]][[1]]
```

```
 MSerror Df    Mean       CV  t.value      LSD
 153.5509 18 122.6667 10.10182 2.100922 10.62822


$Yield[[9]][[2]]
  dependent.var groups
p0     124.8333     a
p1     120.5000     a


$Yield[[10]]
[1] "The means of levels of interaction between two factors are same so dont go for any multiple comparison test"


$Yield[[11]]
$Yield[[11]][[1]]
 MSerror Df    Mean       CV  t.value      LSD
 153.5509 18 122.6667 10.10182 2.100922 15.03057


$Yield[[11]][[2]]
      dependent.var groups
n1:p0      127.1667     a
n0:p0      122.5000     a
n0:p1      120.8333     a
n1:p1      120.1667     a
```

## Pooled RBD Analysis Using R

For conducting pooled RBD analysis across multiple environments, as well as individual RBD analysis for each environment, the "metan" package in R is highly effective. This package is specialized for Multi-Environment Trial Analysis. The data for this analysis, whether in Excel or CSV format, should be organized appropriately to be compatible with the "metan" package's requirements.

## Here, its data arrangement is described in Excel or CSV file

| | ENV | GEN | REP | TA | SI | PH | EH | EPP | EL | KPR | EG | KRP | NKE | Pro | TSS | Car | GCY |
|---|---|---|---|---|---|---|---|---|---|---|---|---|---|---|---|---|---|
| 2 | Anand | 1 | 1 | 69 | 77 | 119 | 51.37 | 1 | 12.66 | 31 | 12.67 | 13.33 | 413.33 | 16.3 | 2.3 | 2.35 | 65 |
| 3 | Anand | 2 | 1 | 65 | 72 | 109.2 | 29.43 | 1 | 14.2 | 24 | 10.75 | 14 | 336 | 17.21 | 1.6 | 2 | 62.5 |
| 4 | Anand | 3 | 1 | 70 | 76 | 124.2 | 50.13 | 1 | 13.5 | 35.3 | 11.83 | 15.36 | 542.21 | 16.26 | 1.7 | 6.35 | 50 |
| 5 | Anand | 4 | 1 | 71 | 79 | 141.6 | 83.15 | 1 | 17.3 | 32.65 | 14 | 11.53 | 376.45 | 15.9 | 1.85 | 2.37 | 136.67 |
| 6 | Anand | 5 | 1 | 71 | 77 | 158 | 70.15 | 1 | 17.5 | 35.36 | 13.25 | 14 | 495.04 | 12.6 | 1 | 4.63 | 100 |
| 7 | Anand | 6 | 1 | 70 | 75 | 95.4 | 25.84 | 1.67 | 10.33 | 29.65 | 12.33 | 14 | 415.1 | 13.02 | 2.4 | 9.33 | 69.44 |
| 8 | Anand | 7 | 1 | 61 | 67 | 90.2 | 62.3 | 2 | 13.67 | 29.33 | 12.33 | 13.33 | 391.11 | 13.62 | 2.36 | 3.65 | 150 |
| 9 | Anand | 8 | 1 | 75 | 80 | 86.6 | 42.38 | 1.5 | 15 | 23.2 | 13 | 12.36 | 286.75 | 15.26 | 1.75 | 3.14 | 44.74 |
| 10 | Anand | 9 | 1 | 73 | 80 | 129.8 | 35.81 | 1.67 | 15.5 | 33 | 10 | 18 | 594 | 14.32 | 2.63 | 7.32 | 125 |
| 11 | Anand | 10 | 1 | 70 | 77 | 81.28 | 24.33 | 1 | 15.64 | 32.45 | 11 | 13 | 421.85 | 13.62 | 1.7 | 5.33 | 25 |
| 88 | Anand | 42 | 2 | 74 | 81 | 100.76 | 32.18 | 1.33 | 13.33 | 27.67 | 13.33 | 14 | 387.33 | 12.05 | 1.43 | 14.26 | 110.83 |
| 89 | Anand | 43 | 2 | 74 | 81 | 108.5 | 40.78 | 1 | 14.33 | 29 | 12 | 14 | 406 | 12.34 | 2.64 | 9.36 | 75 |
| 90 | Anand | 44 | 2 | 71 | 79 | 126.45 | 32.54 | 1.66 | 14.33 | 28.64 | 11 | 13.33 | 381.87 | 14.28 | 2.62 | 6.58 | 83 |
| 91 | Anand | 45 | 2 | 75 | 79 | 123.75 | 55.47 | 1.66 | 13.75 | 27 | 14.25 | 16 | 432 | 14.45 | 2.34 | 8.45 | 103.75 |
| 92 | Godhra | 1 | 1 | 69 | 76 | 148.3 | 58.65 | 1.66 | 15 | 25.5 | 11 | 12.12 | 309.03 | 15.8 | 2.35 | 3.45 | 52.16 |
| 93 | Godhra | 2 | 1 | 62 | 68 | 191.6 | 76.5 | 1.33 | 18 | 32.25 | 12.33 | 13.49 | 434.92 | 16.92 | 1.68 | 2.37 | 96.16 |
| 94 | Godhra | 3 | 1 | 67 | 74 | 139.65 | 74.56 | 1 | 14.63 | 27.75 | 12 | 15.58 | 432.49 | 17.45 | 1.8 | 6.33 | 73.34 |
| 95 | Godhra | 4 | 1 | 68 | 73 | 175.65 | 64.56 | 1.25 | 16 | 28.18 | 14.5 | 13.12 | 369.74 | 16.02 | 1.94 | 4.24 | 138.94 |
| 96 | Godhra | 5 | 1 | 68 | 74 | 95.6 | 45.62 | 1.33 | 13.64 | 24.25 | 9 | 11.13 | 269.99 | 12.12 | 0.99 | 4.88 | 46.87 |
| 97 | Godhra | 6 | 1 | 65 | 73 | 200.1 | 76.56 | 1.66 | 16.89 | 32.15 | 10.76 | 13.92 | 447.41 | 13.05 | 2.55 | 8.67 | 87.5 |
| 98 | Godhra | 7 | 1 | 59 | 66 | 179 | 60.55 | 1.33 | 18.89 | 37.24 | 13 | 14.13 | 526.22 | 13.5 | 1.96 | 3.65 | 127.11 |
| 99 | Godhra | 8 | 1 | 78 | 84 | 135 | 75.65 | 1 | 14.25 | 23.56 | 11 | 12.94 | 304.89 | 16.25 | 1.69 | 3.15 | 49.78 |
| 100 | Godhra | 9 | 1 | 71 | 78 | 193 | 92.56 | 1.3 | 18.46 | 34.08 | 12.33 | 11.6 | 395.24 | 15.02 | 1.21 | 6.95 | 135.74 |
| 101 | Godhra | 10 | 1 | 71 | 76 | 170.94 | 106.5 | 1.32 | 19 | 33.94 | 11 | 11.44 | 388.15 | 13.48 | 1.69 | 7.28 | 162.33 |
| 178 | Godhra | 42 | 2 | 71 | 77 | 158.6 | 86.7 | 1 | 18 | 28 | 13.37 | 13.33 | 373.33 | 12.99 | 1.5 | 13.37 | 142.76 |
| 179 | Godhra | 43 | 2 | 72 | 78 | 156.6 | 94.6 | 1.33 | 16.6 | 31.33 | 13.8 | 13.33 | 417.78 | 13.54 | 2.8 | 13.54 | 159.6 |
| 180 | Godhra | 44 | 2 | 70 | 76 | 161.5 | 58.69 | 1 | 17.83 | 35 | 12.4 | 13.33 | 466.67 | 14.9 | 1.94 | 9.26 | 90 |
| 181 | Godhra | 45 | 2 | 69 | 74 | 178.6 | 94.65 | 1 | 13.83 | 30 | 14.5 | 13.5 | 405 | 15.63 | 1.42 | 9.93 | 159.77 |
| 182 | Sansoli | 1 | 1 | 70 | 75 | 123.65 | 48.9 | 1.5 | 13.83 | 23.51 | 10.14 | 11.17 | 262.7 | 16.05 | 2.33 | 3.17 | 44.34 |
| 183 | Sansoli | 2 | 1 | 66 | 70 | 140.35 | 56.04 | 1.67 | 16.1 | 28.85 | 11.03 | 12.06 | 347.95 | 17.07 | 1.6 | 2.58 | 76.93 |
| 184 | Sansoli | 3 | 1 | 69 | 75 | 121.42 | 64.83 | 1.5 | 14.07 | 26.68 | 11.54 | 14.98 | 399.73 | 16.86 | 1.73 | 6.55 | 67.78 |
| 185 | Sansoli | 4 | 1 | 72 | 78 | 148.69 | 54.65 | 1.13 | 16.65 | 29.33 | 15.09 | 13.65 | 400.39 | 15.96 | 1.92 | 3.98 | 150.46 |
| 186 | Sansoli | 5 | 1 | 71 | 79 | 128.3 | 66.67 | 1.17 | 15.57 | 27.68 | 10.27 | 12.71 | 351.8 | 12.36 | 1.1 | 4.99 | 61.08 |
| 187 | Sansoli | 6 | 1 | 68 | 75 | 113.6 | 50.57 | 1.68 | 13.61 | 25.91 | 8.67 | 11.22 | 290.58 | 13.04 | 2.48 | 8.84 | 56.83 |
| 188 | Sansoli | 7 | 1 | 60 | 68 | 155.5 | 52.6 | 1.5 | 16.28 | 32.09 | 11.2 | 12.18 | 390.77 | 13.56 | 2.16 | 4.32 | 94.39 |
| 189 | Sansoli | 8 | 1 | 76 | 81 | 110.8 | 62.09 | 1.25 | 14.63 | 24.18 | 11.29 | 13.28 | 321.15 | 15.76 | 1.69 | 3.15 | 52.43 |
| 190 | Sansoli | 9 | 1 | 75 | 81 | 161.4 | 77.41 | 1.32 | 16.98 | 31.35 | 11.34 | 10.67 | 334.41 | 14.67 | 1.92 | 7.06 | 114.85 |
| 191 | Sansoli | 10 | 1 | 69 | 75 | 116.3 | 75.99 | 1.16 | 17.32 | 30.94 | 10.03 | 10.42 | 322.54 | 13.55 | 1.7 | 7.04 | 134.89 |

In this dataset, data from three different environments—Anand, Godhra, and Sansoli are presented. It involves testing 45 genotypes across various characteristics using a randomized complete block design (RCBD) with two replications. The dataset's structure includes:

**Column A: ENV:** Indicates the name of the environments.

**Column B: GEN:** Indicates Genotype number.

**Column C: REP:** Indicates Replication number.

**Column D to Q:** Indicates the name of the variable.

In the dataset, rows 2–46 contain data for the first replication of the 45 genotypes from the Anand location, while rows 47–91 hold data for the second replication of these genotypes, also from Anand. Following this pattern, the dataset then includes data for both replications of each genotype from the Godhra and Sansoli locations, respectively.

This arrangement is a standard format for analyzing various variables in Excel. After organizing the data, save the file in .csv format. Then, import it into RStudio, assigning it to a suitably named subset for analysis.

The dataset, once read into the subset in RStudio, will display as follows in the console.

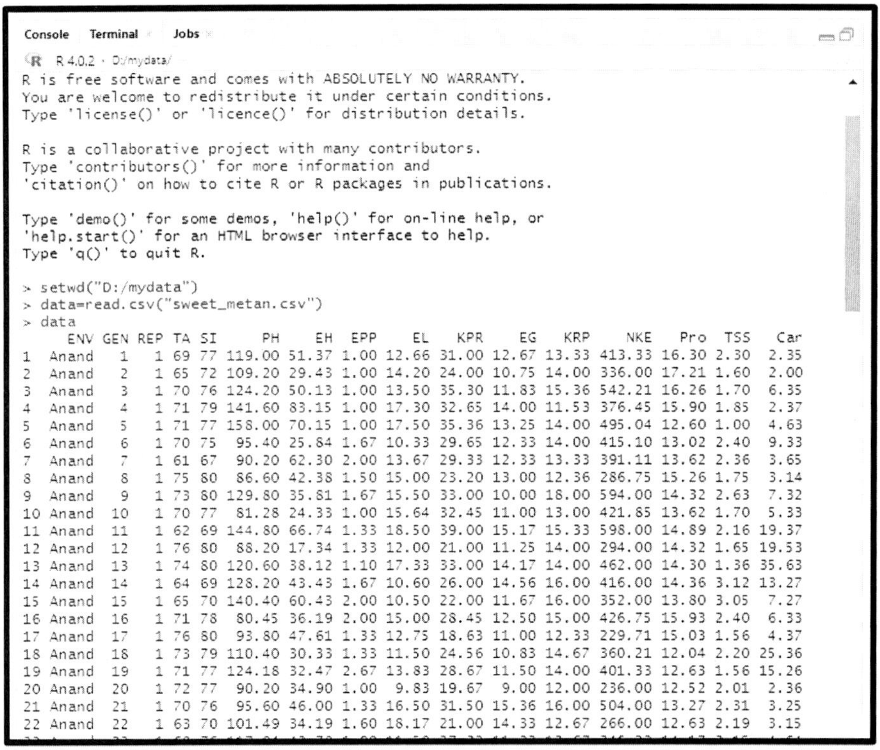

Console Terminal Jobs
R 4.0.2 · D:/mydata/
R is free software and comes with ABSOLUTELY NO WARRANTY.
You are welcome to redistribute it under certain conditions.
Type 'license()' or 'licence()' for distribution details.
R is a collaborative project with many contributors.
Type 'contributors()' for more information and
'citation()' on how to cite R or R packages in publications.
Type 'demo()' for some demos, 'help()' for on-line help, or
'help.start()' for an HTML browser interface to help.
Type 'q()' to quit R.
> setwd("D:/mydata")
> data=read.csv("sweet_metan.csv")
> data
ENV GEN REP TA SI PH EH EPP EL KPR EG KRP NKE Pro TSS Car
1 Anand 1 1 69 77 119.00 51.37 1.00 12.66 31.00 12.67 13.33 413.33 16.30 2.30 2.35
2 Anand 2 1 65 72 109.20 29.43 1.00 14.20 24.00 10.75 14.00 336.00 17.21 1.60 2.00
3 Anand 3 1 70 76 124.20 50.13 1.00 13.50 35.30 11.83 15.36 542.21 16.26 1.70 6.35
4 Anand 4 1 71 79 141.60 83.15 1.00 17.30 32.65 14.00 11.53 376.45 15.90 1.85 2.37
5 Anand 5 1 71 77 158.00 70.15 1.00 17.50 35.36 13.25 14.00 495.04 12.60 1.00 4.63
6 Anand 6 1 70 75 95.40 25.84 1.67 10.33 29.65 12.33 14.00 415.10 13.02 2.40 9.33
7 Anand 7 1 61 67 90.20 62.30 2.00 13.67 29.33 12.33 13.33 391.11 13.62 2.36 3.65
8 Anand 8 1 75 80 86.60 42.38 1.50 15.00 23.20 13.00 12.36 286.75 15.26 1.75 3.14
9 Anand 9 1 73 80 129.80 35.81 1.67 15.50 33.00 10.00 18.00 594.00 14.32 2.63 7.32
10 Anand 10 1 70 77 81.28 24.33 1.00 15.64 32.45 11.00 13.00 421.85 13.62 1.70 5.33
11 Anand 11 1 62 69 144.80 66.74 1.33 18.50 39.00 15.17 15.33 598.00 14.89 2.16 19.37
12 Anand 12 1 76 80 88.20 17.34 1.33 12.00 21.00 11.25 14.00 294.00 14.32 1.65 19.53
13 Anand 13 1 74 80 120.60 38.12 1.10 17.33 33.00 14.17 14.00 462.00 14.30 1.36 35.63
14 Anand 14 1 64 69 128.20 43.43 1.67 10.60 26.00 14.56 16.00 416.00 14.36 3.12 13.27
15 Anand 15 1 65 70 140.40 60.43 2.00 10.50 22.00 11.67 16.00 352.00 13.80 3.05 7.27
16 Anand 16 1 71 78 80.45 36.19 2.00 15.00 28.45 12.50 15.00 426.75 15.93 2.40 6.33
17 Anand 17 1 76 80 93.80 47.61 1.33 12.75 18.63 11.00 12.33 229.71 15.03 1.56 4.37
18 Anand 18 1 73 79 110.40 30.33 1.33 11.50 24.56 10.83 14.67 360.21 12.04 2.20 25.36
19 Anand 19 1 71 77 124.18 32.47 2.67 13.83 28.67 11.50 14.00 401.33 12.63 1.56 15.26
20 Anand 20 1 72 77 90.20 34.90 1.00 9.83 19.67 9.00 12.00 236.00 12.52 2.01 2.36
21 Anand 21 1 70 76 95.60 46.00 1.33 16.50 31.50 15.36 16.00 504.00 13.27 2.31 3.25
22 Anand 22 1 63 70 101.49 34.19 1.60 18.17 21.00 14.33 12.67 266.00 12.63 2.19 3.15

55 116.67
56 165.00
57 55.42
58 170.63
[ reached 'max' / getOption("max.print") -- omitted 212 rows ]
Type here to search

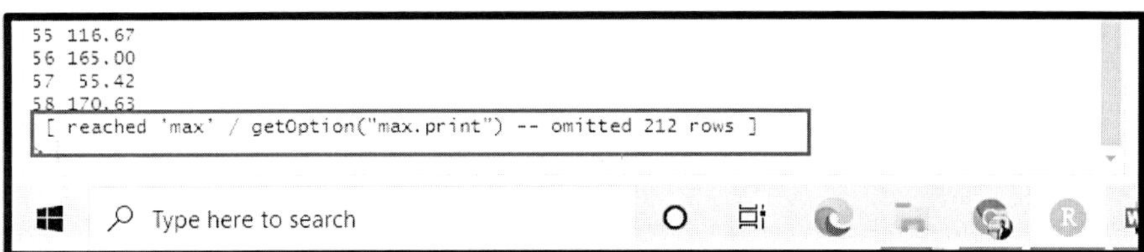

In cases where the dataset is too large to be fully displayed in the RStudio console, leading to a notification like "[reached 'max'/getOption('max. Print')—omitted 212 rows]."

You can adjust the settings to view the entire dataset. To do this, increase the maximum number of lines that can be printed in the console by using the options function with the **"max.print"** parameter.

### Code

```
options(max.print = 100000)
```

After running this code, input the name of the data subset into the script editor and execute it. This will display the full dataset in the console.

```
RStudio
File Edit Code View Plots Session Build Debug Profile Tools Help
Go to file/function   Addins
Untitled1*
Source on Save
setwd("D:/mydata")
data=read.csv("sweet_metan.csv")

options(max.print = 100000)
data
```

### Result in console

```
Console  Terminal  Jobs
R 4.0.2 · D:/mydata/
58 170.63
 [ reached 'max' / getOption("max.print") -- omitted 212 rows ]
> options(max.print = 100000)
> data
        ENV GEN REP TA SI     PH    EH  EPP    EL   KPR    EG   KRP    NKE   Pro  TSS
1     Anand   1   1 69 77 119.00 51.37 1.00 12.66 31.00 12.67 13.33 413.33 16.30 2.30
2     Anand   2   1 65 72 109.20 29.43 1.00 14.20 24.00 10.75 14.00 336.00 17.21 1.60
3     Anand   3   1 70 76 124.20 50.13 1.00 13.50 35.30 11.83 15.36 542.21 16.26 1.70
4     Anand   4   1 71 79 141.60 83.15 1.00 17.30 32.65 14.00 11.53 376.45 15.90 1.85
5     Anand   5   1 71 77 158.00 70.15 1.00 17.50 35.36 13.25 14.00 495.04 12.60 1.00
6     Anand   6   1 70 75  95.40 25.84 1.67 10.33 29.65 12.33 14.00 415.10 13.02 2.40
7     Anand   7   1 61 67  90.20 62.30 2.00 13.67 29.33 12.33 13.33 391.11 13.62 2.36
8     Anand   8   1 75 80  86.60 42.38 1.50 15.00 23.20 13.00 12.36 286.75 15.26 1.75
9     Anand   9   1 73 80 129.80 35.81 1.67 15.50 33.00 10.00 18.00 594.00 14.32 2.63
10    Anand  10   1 70 77  81.28 24.33 1.00 15.64 32.45 11.00 13.00 421.85 13.62 1.70
11    Anand  11   1 62 69 144.80 66.74 1.33 18.50 39.00 15.17 15.33 598.00 14.89 2.16
12    Anand  12   1 76 80  88.20 17.34 1.33 12.00 21.00 11.25 14.00 294.00 14.32 1.65
13    Anand  13   1 74 80 120.60 38.12 1.10 17.33 33.00 14.17 14.00 462.00 14.30 1.36
14    Anand  14   1 64 69 128.20 43.43 1.67 10.60 26.00 14.56 16.00 416.00 14.36 3.12
15    Anand  15   1 65 70 140.40 60.43 2.00 10.50 22.00 11.67 16.00 352.00 13.80 3.05
```

```
261  6.35   95.81
262 15.35  135.87
263  3.33   15.22
264 14.33   96.27
265 15.33   30.73
266  4.25   85.27
267 14.33   94.11
268 10.00  170.30
269  6.33   80.48
270  6.54  112.70
>
```

Type here to search

By adjusting the maximum print limit in RStudio, it becomes possible to view all rows of the dataset, numbered from 1 to 270. This dataset comprehensively includes data collected from Anand, Godhra, and Sansoli, formatted appropriately for analysis.

Next, initiate the "metan" package in RStudio for further analysis.

Once this package is launched, the console will display a confirmation of its activation.

It will look like this in console.

Console Terminal Jobs

R 4.0.2 · D:/mydata/

```
> library("metan")
Registered S3 method overwritten by 'GGally':
  method from
  +.gg   ggplot2
|==========================================================|
| Multi-Environment Trial Analysis (metan) v1.15.0         |
| Author: Tiago Olivoto                                    |
| Type 'citation('metan')' to know how to cite metan       |
| Type 'vignette('metan_start')' for a short tutorial      |
| Visit 'https://bit.ly/pkgmetan' for a complete tutorial  |
|==========================================================|
Warning message:
package 'metan' was built under R version 4.0.5
> |
```

**This metan package contains the code for ANOVA and pooled ANOVA analysis**

i. **Perform analysis of variance for each environment**

**Code**

```
ind_an=anova_ind(data,
        env = ENV,
        gen = GEN,
        rep = REP,
        resp = c(TA, GCY))
```

**In the script editor**

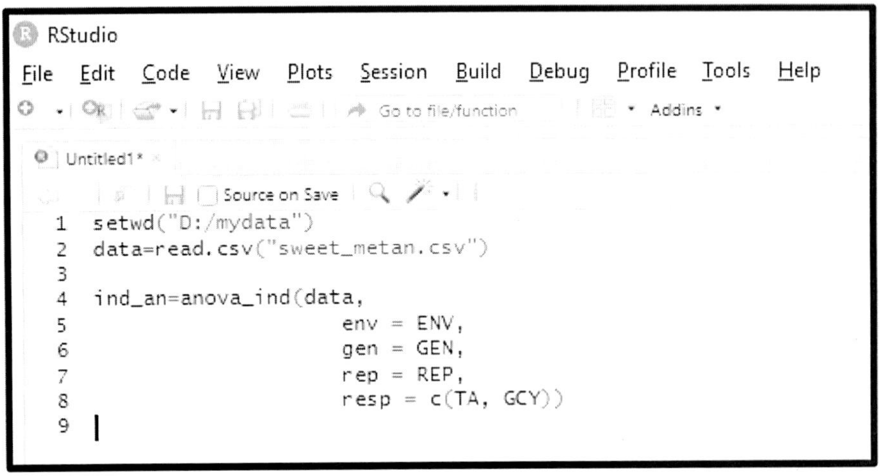

RStudio

File Edit Code View Plots Session Build Debug Profile Tools Help

Go to file/function · Addins

Untitled1*

Source on Save

```
setwd("D:/mydata")
data=read.csv("sweet_metan.csv")

ind_an=anova_ind(data,
                 env = ENV,
                 gen = GEN,
                 rep = REP,
                 resp = c(TA, GCY))

```

This function conducts a within-environment analysis of variance for data collected in randomized complete block or alpha-lattice designs. It returns key statistical values including mean squares, *p*-values, coefficient of variation, heritability, and accuracy of selection. Here are the arguments explained:

- **data:** The dataset containing the columns related to environments, genotypes, replication/block, and response variable(s).
- **env:** Specifies the column containing the environment levels.
- **gen:** Indicates the column with the genotype levels.
- **rep:** Refers to the column with replication or block levels.

- **resp:** The response variable(s). To analyze multiple variables in a single procedure a vector of variables may be used. For example resp = c(var1, var2, var3).
- **block:** Defaults to NULL. In this case, a randomized complete block design is considered. If the block is informed, then a resolvable alpha-lattice design is employed. All effects, except the error, are assumed to be fixed.
- **verbose:** A logical parameter. When set to FALSE, the function operates without displaying the process details. Result in the console.

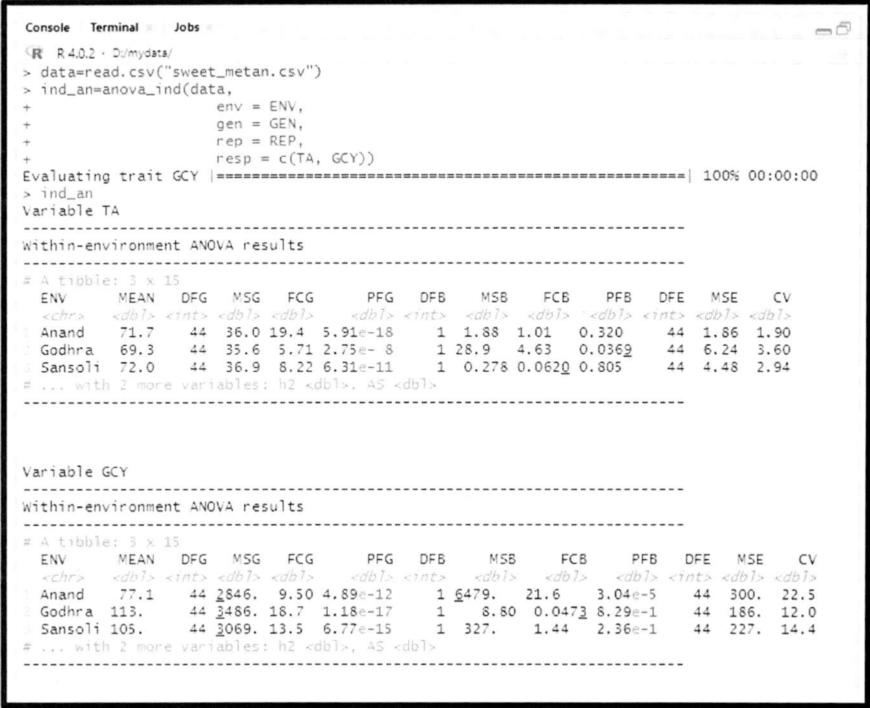

```
Console   Terminal   Jobs
R 4.0.2 · D:/mydata/
> data=read.csv("sweet_metan.csv")
> ind_an=anova_ind(data,
+                    env = ENV,
+                    gen = GEN,
+                    rep = REP,
+                    resp = c(TA, GCY))
Evaluating trait GCY |====================================================| 100% 00:00:00
> ind_an
Variable TA
---------------------------------------------------------------------------
Within-environment ANOVA results
---------------------------------------------------------------------------
# A tibble: 3 x 15
  ENV      MEAN   DFG   MSG   FCG      PFG   DFB    MSB    FCB     PFB   DFE   MSE    CV
  <chr>   <dbl> <int> <dbl> <dbl>    <dbl> <int>  <dbl>  <dbl>   <dbl> <int> <dbl> <dbl>
1 Anand    71.7    44  36.0 19.4  5.91e-18     1  1.88  1.01    0.320     44  1.86  1.90
2 Godhra   69.3    44  35.6  5.71 2.75e- 8     1 28.9   4.63    0.0369    44  6.24  3.60
3 Sansoli  72.0    44  36.9  8.22 6.31e-11     1  0.278 0.0620  0.805     44  4.48  2.94
# ... with 2 more variables: h2 <dbl>, AS <dbl>
---------------------------------------------------------------------------


Variable GCY
---------------------------------------------------------------------------
Within-environment ANOVA results
---------------------------------------------------------------------------
# A tibble: 3 x 15
  ENV      MEAN   DFG   MSG   FCG      PFG   DFB    MSB     FCB      PFB   DFE   MSE    CV
  <chr>   <dbl> <int> <dbl> <dbl>    <dbl> <int>  <dbl>   <dbl>    <dbl> <int> <dbl> <dbl>
1 Anand    77.1    44 2846.  9.50 4.89e-12     1 6479.   21.6    3.04e-5    44  300.  22.5
2 Godhra  113.     44 3486. 18.7  1.18e-17     1    8.80  0.0473 8.29e-1    44  186.  12.0
3 Sansoli 105.     44 3069. 13.5  6.77e-15     1  327.    1.44   2.36e-1    44  227.  14.4
# ... with 2 more variables: h2 <dbl>, AS <dbl>
---------------------------------------------------------------------------
```

To print the results of the analysis in a Word document, the "bruceR" package in R can be utilized. Before proceeding, it's important to locate where the ANOVA results are stored within RStudio. Check the RStudio environment for the subset named "ind_an" which contains the ANOVA results. Once confirmed, you can proceed to export these results to a Word file using functionalities provided by the "bruceR" package.

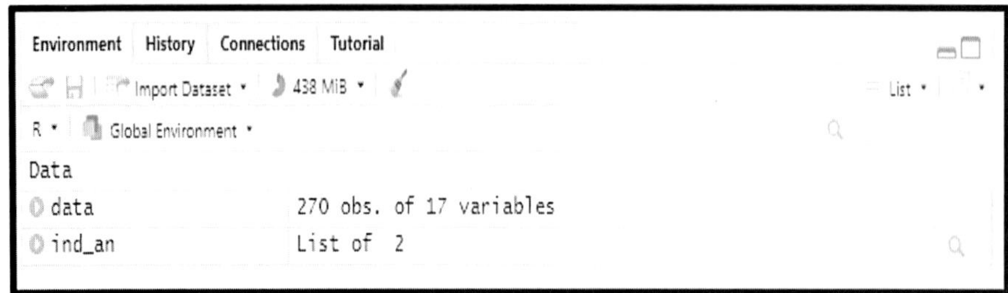


To view the script, click on "ind_an" and then look at the script editor.

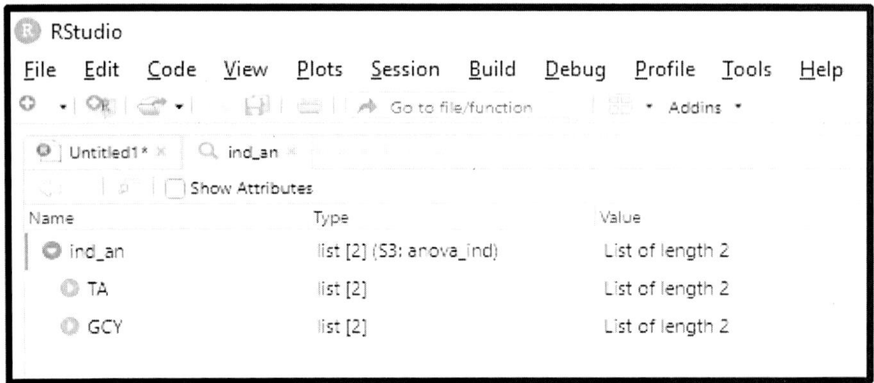


In this panel, the ANOVA for TA and GCY is calculated and the results are stored. Please click on the blue button labeled "TA" to proceed.

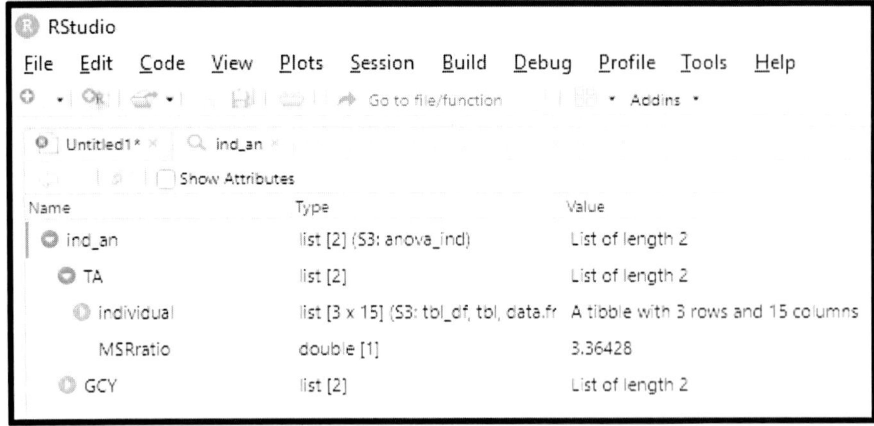


Upon further exploration in the RStudio environment, two additional options become visible: **"individual**" and "**MSRatio**." The **"MSRatio"** displays a value of 3.3642, which represents the ratio between the highest and lowest residual mean squares.

When you place the cursor over "**individual**," two new buttons appear, as highlighted in a square in the interface. Alongside these buttons, a dialogue box is also visible, showing a tibble (a type of data frame in R) that contains 3 rows and 15 columns. This tibble likely presents detailed data or results relevant to the individual analysis being conducted.

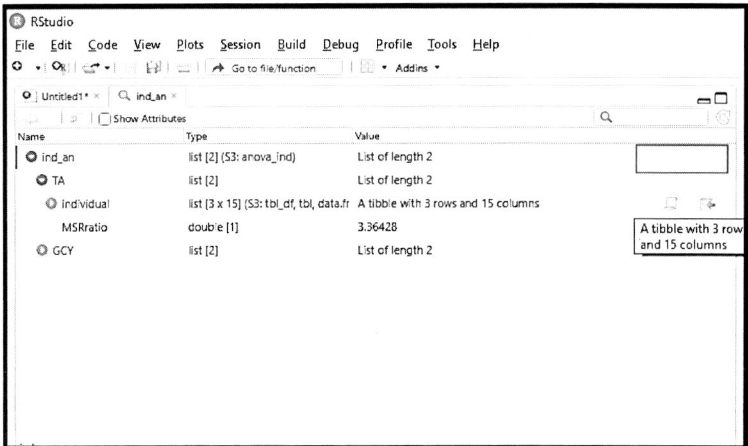


Then click on the paper button

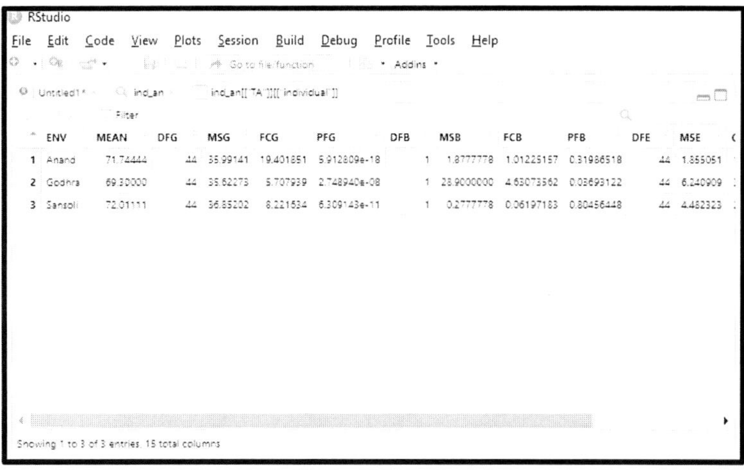


| | ENV | MEAN | DFG | MSG | FCG | PFG | DFB | MSB | FCB | PFB | DFE | MSE |
|---|---|---|---|---|---|---|---|---|---|---|---|---|
| 1 | Anand | 71.74444 | 44 | 35.99141 | 19.401851 | 5.912809e-18 | 1 | 1.8777778 | 1.01225157 | 0.31986518 | 44 | 1.855051 |
| 2 | Godhra | 69.30000 | 44 | 35.62273 | 5.707939 | 2.748940e-08 | 1 | 28.9000000 | 4.63073562 | 0.03693122 | 44 | 6.240909 |
| 3 | Sansoli | 72.01111 | 44 | 36.85202 | 8.221534 | 6.309143e-11 | 1 | 0.2777778 | 0.06197183 | 0.80456448 | 44 | 4.482323 |

This table will appear immediately after clicking on that button.

Information regarding table is given at the bottom of the script editor.

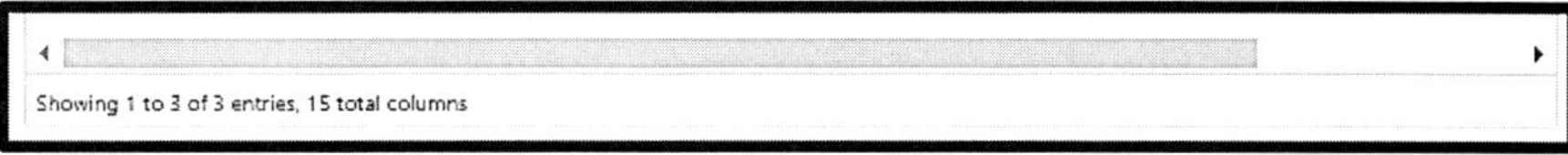

This table displays the individual ANOVA results for our dataset, which are saved in the "ind_an" subset, including the TA and individual variables.

The title of the table indicates this information.

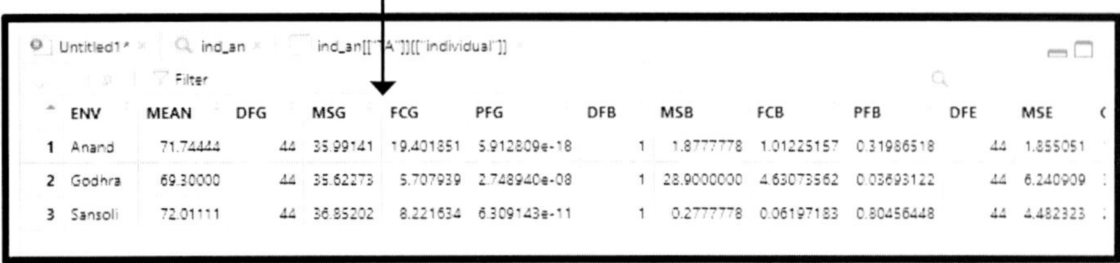

| | ENV | MEAN | DFG | MSG | FCG | PFG | DFB | MSB | FCB | PFB | DFE | MSE |
|---|---|---|---|---|---|---|---|---|---|---|---|---|
| 1 | Anand | 71.74444 | 44 | 35.99141 | 19.401851 | 5.912809e-18 | 1 | 1.8777778 | 1.01225157 | 0.31986518 | 44 | 1.855051 |
| 2 | Godhra | 69.30000 | 44 | 35.62273 | 5.707939 | 2.748940e-08 | 1 | 28.9000000 | 4.63073562 | 0.03693122 | 44 | 6.240909 |
| 3 | Sansoli | 72.01111 | 44 | 36.85202 | 8.221634 | 6.309143e-11 | 1 | 0.2777778 | 0.06197183 | 0.80456448 | 44 | 4.482323 |

Now, launch the "bruceR" package and type the code given below for printing this table

**Code**

```
print_table(ind_an$TA$individual, file="ANOVA of TA")
```

And run

```
Console   Terminal   Jobs
R 4.0.2 · D:/mydata/
> print_table(ind_an$TA$individual, file="ANOVA of TA")
✓ Table saved to "D:/mydata/ANOVA of TA.doc"

>
```

The console indicates that the results of the individual ANOVA have been saved in the working directory as a document file, named "ANOVA of TA.doc." To review the details and findings of the ANOVA analysis, open and examine the contents of this .doc file.

| | ENV | MEAN | DFG | MSG | FCG | PFG | DFB | MSB | FCB | PFB | DFE | MSE | CV | h2 | AS |
|---|---|---|---|---|---|---|---|---|---|---|---|---|---|---|---|
| 1 | Anand | 71.744 | 44.000 | 35.991 | 19.402 | 0.000 | 1.000 | 1.878 | 1.012 | 0.320 | 44.000 | 1.855 | 1.898 | 0.948 | 0.974 |
| 2 | Godhra | 69.300 | 44.000 | 35.623 | 5.708 | 0.000 | 1.000 | 28.900 | 4.631 | 0.037 | 44.000 | 6.241 | 3.605 | 0.825 | 0.908 |
| 3 | Sansoli | 72.011 | 44.000 | 36.852 | 8.222 | 0.000 | 1.000 | 0.278 | 0.062 | 0.805 | 44.000 | 4.482 | 2.940 | 0.878 | 0.937 |

### *Description of Output*

The output from the analysis is a list, where each element corresponds to the results for one variable. The structure of this output varies based on the experimental design:

**For RCBD**

- **MEAN:** The grand mean.
- **DFG, DFB, and DFE:** The degree of freedom for genotypes, blocks, and error, respectively.
- **MSG, MSB, and MSE:** The mean squares for genotype blocks, and error, respectively.
- **FCG and FCB:** The *F*-calculated for genotype and blocks, respectively.
- **PFG and PFB:** The *p*-values for genotype and blocks, respectively.
- **CV:** coefficient of variation.
- **h2:** broad-sense heritability.
- **AS:** accuracy of selection (square root of h2).

**For analysis in alpha-lattice designs**

- **MEAN:** The grand mean.
- **DFG, DFCR, and DFIB_R, and DFE:** The degree of freedom for genotypes, complete replicates, incomplete blocks within replicates, and error, respectively.
- **MSG, MSCR, MSIB_R:** The mean squares for genotype, replicates, incomplete blocks within replicates, and error, respectively.
- **FCG, FCR, FCIB_R:** The F-calculated for genotype, replicates, and incomplete blocks within replicates, respectively.
- **PFG, PFCR, PFIB_R:** The P-values for genotype, replicates, and incomplete blocks within replicates, respectively.
- **CV:** coefficient of variation.
- **h2:** broad-sense heritability.
- **AS:** accuracy of selection (square root of h2).

If we want to get mean of the **individual environment,** write down the code below:

Code: `get_model_data(ind_an)`

**Console**

```
> get_model_data(ind_an)
Class of the model: anova_ind
Variable extracted: MEAN
# A tibble: 3 x 3
  ENV        TA    GCY
  <chr>    <dbl> <dbl>
1 Anand     71.7  77.1
2 Godhra    69.3 113.
3 Sansoli   72.0 105.
```

For *p*-value of genotypic effects

Code: get_model_data(ind_an, "PFG")

**Console**

```
> get_model_data(ind_an, "PFG")
Class of the model: anova_ind
Variable extracted: PFG
# A tibble: 3 x 3
  ENV            TA      GCY
  <chr>       <dbl>    <dbl>
1 Anand    5.91e-18 4.89e-12
2 Godhra   2.75e- 8 1.18e-17
3 Sansoli  6.31e-11 6.77e-15
>
```

ii. **Perform pooled analysis of variance**

**Code:**

j_an = anova_joint(data_ge, env = ENV, gen = GEN,rep = REP, resp = c(TA, GCY))

```
RStudio
File Edit Code View Plots Session Build Debug Profile Tools Help
Go to file/function   Addins
Untitled1*
Source on Save
setwd("D:/mydata")
data=read.csv("sweet_metan.csv")

j_an <- anova_joint(data,
                    env = ENV,
                    gen = GEN,
                    rep = REP,
                    resp = c(TA, GCY))

```

This function conducts a combined analysis of variance, evaluating genotype-vs-environment interactions using both Randomized Complete Block and Alpha-Lattice designs.

## Description of code

- **data:** This is the dataset which should include columns for environments, genotypes, replications/blocks, and the response variable(s).
- **env, gen and rep:** Specifies the different environment, genotype and replications.
- **resp:** The response variable(s). To analyze multiple variables in a single procedure a vector of variables may be used. For example resp = c(var1, var2, var3).
- **block:** Defaults to NULL. In this case, a randomized complete block design is considered. If block is informed, then a resolvable alpha-lattice design is employed. All effects, except the error, are assumed to be fixed.

## Output in console

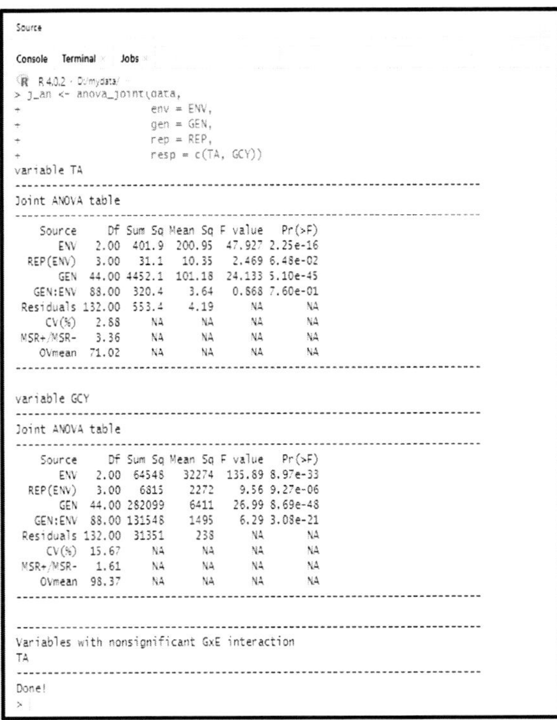

```
Source

Console  Terminal  Jobs

R 4.0.2 · D:/mydata/
> j_an <- anova_joint(data,
+                     env = ENV,
+                     gen = GEN,
+                     rep = REP,
+                     resp = c(TA, GCY))
variable TA
---------------------------------------------------------------------------
Joint ANOVA table
---------------------------------------------------------------------------
    Source     Df Sum Sq Mean Sq F value   Pr(>F)
       ENV   2.00  401.9  200.95  47.927 2.25e-16
  REP(ENV)   3.00   31.1   10.35   2.469 6.48e-02
       GEN  44.00 4452.1  101.18  24.133 5.10e-45
   GEN:ENV  88.00  320.4    3.64   0.868 7.60e-01
 Residuals 132.00  553.4    4.19      NA       NA
     CV(%)   2.88     NA      NA      NA       NA
 MSR+/MSR-   3.36     NA      NA      NA       NA
    OVmean  71.02     NA      NA      NA       NA
---------------------------------------------------------------------------

variable GCY
---------------------------------------------------------------------------
Joint ANOVA table
---------------------------------------------------------------------------
    Source     Df Sum Sq Mean Sq F value   Pr(>F)
       ENV   2.00  64548   32274  135.89 8.97e-33
  REP(ENV)   3.00   6815    2272    9.56 9.27e-06
       GEN  44.00 282099    6411   26.99 8.69e-48
   GEN:ENV  88.00 131548    1495    6.29 3.08e-21
 Residuals 132.00  31351     238      NA       NA
     CV(%)  15.67     NA      NA      NA       NA
 MSR+/MSR-   1.61     NA      NA      NA       NA
    OVmean  98.37     NA      NA      NA       NA
---------------------------------------------------------------------------

---------------------------------------------------------------------------
Variables with nonsignificant GxE interaction
TA
---------------------------------------------------------------------------
Done!
>
```

Right after completing the analysis, the program identifies and displays the names of variables with non-significant Genotype-Environment Interaction (GEI).

To obtain more detailed output for each variable, you can enter the name of the specific subset in the script editor and execute it. For example, typing "j_an" and running this command will produce an expanded set of results in the console, pertaining to each variable in question.

```
Console   Terminal ×   Jobs ×
R 4.0.2 · D:/mydata/
> j_an
Variable TA
---------------------------------------------------------------------------
$anova
     Source         Df      Sum Sq    Mean Sq    F value        Pr(>F)
1       ENV   2.000000  401.89630 200.948148 47.9274041 2.251326e-16
2  REP(ENV)   3.000000   31.05556  10.351852  2.4689821 6.481101e-02
3       GEN  44.000000 4452.07407 101.183502 24.1329050 5.098538e-45
4   GEN:ENV  88.000000  320.43704   3.641330  0.8684802 7.599672e-01
5 Residuals 132.000000  553.44444   4.192761         NA           NA
6     CV(%)   2.883224         NA         NA         NA           NA
7 MSR+/MSR-   3.364280         NA         NA         NA           NA
8    OVmean  71.018519         NA         NA         NA           NA

$model
Call:
   aov(formula = mean ~ GEN + ENV + GEN:ENV + ENV/REP, data = data)

Terms:
                     GEN      ENV  GEN:ENV  ENV:REP Residuals
Sum of Squares  4452.074  401.896  320.437   31.056   553.444
Deg. of Freedom       44        2       88        3       132

Residual standard error: 2.047623
Estimated effects may be unbalanced

$augment
# A tibble: 270 x 11
   ENV   GEN   REP    mean   hat sigma fitted  resid stdres se.fit factors
   <fct> <fct> <fct> <int> <dbl> <dbl>  <dbl>  <dbl>  <dbl>  <dbl> <chr>
 1 Anand 1     1        69 0.511  2.04   70.9 -1.86  -1.30    1.46 1_1
 2 Anand 2     1        65 0.511  2.05   64.4  0.644  0.450   1.46 2_1
 3 Anand 3     1        70 0.511  2.05   68.9  1.14   0.799   1.46 3_1
 4 Anand 4     1        71 0.511  2.05   70.4  0.644  0.450   1.46 4_1
 5 Anand 5     1        71 0.511  2.05   71.9 -0.856 -0.598   1.46 5_1
 6 Anand 6     1        70 0.511  2.05   70.9 -0.856 -0.598   1.46 6_1
 7 Anand 7     1        61 0.511  2.05   60.4  0.644  0.450   1.46 7_1
 8 Anand 8     1        75 0.511  2.05   76.4 -1.36  -0.947   1.46 8_1
 9 Anand 9     1        73 0.511  2.05   73.4 -0.356 -0.248   1.46 9_1
10 Anand 10    1        70 0.511  2.06   69.9  0.144  0.101   1.46 10_1
# ... with 260 more rows
```

Further information regarding pooled ANOVA, akin to what is provided in the individual ANOVA, can also be extracted. The basic output for pooled ANOVA typically includes:

The results of the Pooled ANOVA for TA, along with additional detailed information, are directly provided in a .doc format. These results can be conveniently extracted from RStudio using the "bruceR" package, as outlined in the previous topic.

**Pooled ANOVA**

| | Source | *Df* | Sum sq | Mean sq | *F* | *p* | |
|---|---|---|---|---|---|---|---|
| 1 | ENV | 2.000 | 401.896 | 200.948 | 47.927 | <0.001 | *** |
| 2 | REP(ENV) | 3.000 | 31.056 | 10.352 | 2.469 | 0.065 | . |
| 3 | GEN | 44.000 | 4452.074 | 101.184 | 24.133 | <0.001 | *** |
| 4 | GEN:ENV | 88.000 | 320.437 | 3.641 | 0.868 | 0.760 | |
| 5 | Residuals | 132.000 | 553.444 | 4.193 | | | |
| 6 | CV(%) | 2.883 | | | | | |
| 7 | MSR+/MSR- | 3.364 | | | | | |
| 8 | OVmean | 71.019 | | | | | |

**Details**

| | Parameters | Mean |
|---|---|---|
| 1 | Mean | 71.02 |
| 2 | SE | 0.28 |
| 3 | SD | 4.62 |
| 4 | CV | 6.52 |
| 5 | Min | 57 (2 in Godhra) |
| 6 | Max | 79 (32 in Anand) |
| 7 | MinENV | Godhra (69.3) |
| 8 | MaxENV | Sansoli (72.01) |
| 9 | MinGEN | 7 (60.83) |
| 10 | MaxGEN | 35 (76.83) |

## Augment

| | ENV | GEN | REP | mean | hat | sigma | fitted | resid | stdres | se.fit | factors |
|---|---|---|---|---|---|---|---|---|---|---|---|
| 1 | Anand | 1 | 1 | 69.000 | 0.511 | 2.042 | 70.856 | −1.856 | −1.296 | 1.464 | 1_1 |
| 2 | Anand | 2 | 1 | 65.000 | 0.511 | 2.054 | 64.356 | 0.644 | 0.450 | 1.464 | 2_1 |
| 3 | Anand | 3 | 1 | 70.000 | 0.511 | 2.050 | 68.856 | 1.144 | 0.799 | 1.464 | 3_1 |
| 4 | Anand | 4 | 1 | 71.000 | 0.511 | 2.054 | 70.356 | 0.644 | 0.450 | 1.464 | 4_1 |
| 5 | Anand | 5 | 1 | 71.000 | 0.511 | 2.053 | 71.856 | −0.856 | −0.598 | 1.464 | 5_1 |
| 6 | Anand | 6 | 1 | 70.000 | 0.511 | 2.053 | 70.856 | −0.856 | −0.598 | 1.464 | 6_1 |
| 7 | Anand | 7 | 1 | 61.000 | 0.511 | 2.054 | 60.356 | 0.644 | 0.450 | 1.464 | 7_1 |
| 8 | Anand | 8 | 1 | 75.000 | 0.511 | 2.048 | 76.356 | −1.356 | −0.947 | 1.464 | 8_1 |
| 9 | Anand | 9 | 1 | 73.000 | 0.511 | 2.055 | 73.356 | −0.356 | −0.248 | 1.464 | 9_1 |
| 10 | Anand | 10 | 1 | 70.000 | 0.511 | 2.055 | 69.856 | 0.144 | 0.101 | 1.464 | 10_1 |
| 11 | Anand | 11 | 1 | 62.000 | 0.511 | 2.053 | 62.856 | −0.856 | −0.598 | 1.464 | 11_1 |
| 12 | Anand | 12 | 1 | 76.000 | 0.511 | 2.045 | 74.356 | 1.644 | 1.149 | 1.464 | 12_1 |
| 13 | Anand | 13 | 1 | 74.000 | 0.511 | 2.045 | 72.356 | 1.644 | 1.149 | 1.464 | 13_1 |
| 14 | Anand | 14 | 1 | 64.000 | 0.511 | 2.053 | 64.856 | −0.856 | −0.598 | 1.464 | 14_1 |
| 15 | Anand | 15 | 1 | 65.000 | 0.511 | 2.050 | 63.856 | 1.144 | 0.799 | 1.464 | 15_1 |
| 16 | Anand | 16 | 1 | 71.000 | 0.511 | 2.055 | 71.356 | −0.356 | −0.248 | 1.464 | 16_1 |

### *Description of Output*

The output is organized as a list, with each element representing the results for a particular variable. It includes the following components:

- **ANOVA:** The two-way ANOVA table.
- **model:** Represents model of class lm (linear model).
- **augment:** This section provides detailed information about each observation in the dataset. It includes the predicted values under the "fitted" column, the residuals in "resid," standardized residuals in "stdres," elements of the "hat" matrix in "hat," and standard errors for the predicted values listed in the "se.fit" column.
- **details:** This is a tibble containing specific data points such as:
    - **Ngen:** the number of genotypes;
    - **OVmean:** Grand mean;
    - **Min:** Minimum observed (returning the genotype and replication/ block)
    - **Max**: Maximum observed
    - **MinGEN**: The genotype with the lowest value (loser).
    - **MaxGEN:** The genotype with the highest value (winner).

# 10

# Latin Square Design (LSD)

In this chapter, we explore Latin square design (LSD), a specialized form of experimental design renowned for its efficiency in controlling variation. This chapter thoroughly discovers the foundational principles, structure, and application of LSD in experimental research. It details mathematical models for LSD and guides readers through the interpretation of ANOVA tables. Furthermore, the chapter provides a systematic approach to conducting LSD analysis using R software.

## Latin Square Design

The LSD is a statistical design used extensively in experiments, especially when specific conditions are met:

- Controlled variation in two directions: LSD is particularly suited for experiments where variation can be controlled along two different factors. This design structures the experimental units into a square matrix, allowing two sources of variability to be accounted for.
- Adherence to experimental design principles: Like other designs, LSD follows the key principles of randomization, replication, and local control. Randomization in LSD is applied within the rows and columns of the square matrix, ensuring unbiased treatment comparisons.
- Efficient use of resources: LSD allows for an effective utilization of resources by controlling two variables simultaneously. This makes it a more efficient design compared to RBD or CRD when there are two sources of nuisance variables.

## Layout of LSD

The "layout" pertains to how treatments are arranged within each row and column. In an LSD, the arrangement is such that the number of treatments equals the number of rows and columns. For example, with 'm' treatments, the design will consist of $m^2$ (m-squared) experimental units or plots, organized in a square matrix. Each row and each column in this matrix will have "m" plots. In this setup, the 'm' treatments are randomly distributed across rows and columns, ensuring each treatment appears only once in each row and column. This structure is known as an $m \times m$ LSD and is commonly utilized in agricultural experiments. Typically, the number of treatments in an LSD range from 5 to 12.

### *Layout of Five Treatments and Five Replications*

| A | B | C | D | E |
|---|---|---|---|---|
| B | C | D | E | A |
| C | D | E | A | B |
| D | E | A | B | C |
| E | A | B | C | D |

| A | B | C | D | E |
|---|---|---|---|---|
| B | A | D | E | C |
| C | E | A | B | D |
| D | C | E | A | B |
| E | D | B | C | A |

| A | B | C | D | E |
|---|---|---|---|---|
| B | A | E | C | D |
| C | D | A | E | B |
| D | E | B | A | C |
| E | C | D | B | A |

These squares are called *standard squares* and their randomization procedure in a 5 × 5 LSD.

### *Standard Squares*

- A standard square is an arrangement where the first row and the first column are in alphabetical order.
- Two standard squares are termed conjugate squares if the rows of one are the columns of the other.

## Randomization Procedure for 5 × 5 LSD

- Layout preparation: Divide the experimental material into rows and columns based on the number of treatments.
- Selection of standard square: Choose a t × t standard square from sources like Fisher and Yates' statistical tables.
- Available options: For a 5 × 5 Latin square, there are 56 potential standard squares—25 pairs and 6 individual squares.
- Pair assignment: The first 25 squares are paired (e.g., (1,2), (3,4)), where the first number represents the standard square and the second indicates its conjugate.
- Initial randomization: Select a random number between 1 and 56. If it's 50 or below, determine if it's odd or even. Use the standard square for odd numbers, and the conjugate for even. For numbers between 51 and 56, select the corresponding standard square directly.

- Example selection: If the random number is 48, the corresponding standard square is:

| A | B | C | D | E |
|---|---|---|---|---|
| B | E | A | C | D |
| C | D | B | E | A |
| D | C | E | A | B |
| E | A | D | B | C |

- Conjugate square: The conjugate of this square is:

| A | B | C | D | E |
|---|---|---|---|---|
| B | E | D | C | A |
| C | A | B | E | D |
| D | C | E | A | B |
| E | D | A | B | C |

- Column randomization: Randomize the columns using a set of five random numbers. This changes the order of the columns.
- Row randomization: Randomize the rows (except the first row) using four random numbers ranging from 2 to 5. This rearranges the order of the rows.
- Final layout: The resulting square is the final 5 × 5 LSD (Table 1).

Statistical model for LSD: $y_{ijk} = \mu + \alpha_i + \beta_j + \gamma_k + \varepsilon_{ijk}$,

where, $Y_{ijk}$ = response of yield, $\mu$ = general mean, $\alpha_i$ = effect of *i*th treatment, $\beta_j$ = effect due to *j*th row, $\gamma_k$ = effect due to kth column; $\varepsilon_{ij}$= uncontrolled variation associated with *i*th treatment of *j*th row in *k*th columns.

**Table 1.** Analysis of variance.

| **Source of variation (S.V.)** | **Degrees of freedom (*df*)** | **Sum of square (S.S.)** | **Mean square (M.S.)** | ***F*-calculated value** | ***F*-tabulated value** |
|---|---|---|---|---|---|
| Row | $m - 1$ | Row S.S. | R.M.S. = Row S.S./$m - 1$ | $F_r$ = R.M.S./ M.S.E. | F[(m-1), (m-1)(m-2) at 0.05 level of significance |
| Column | $m - 1$ | Column S.S. | C.M.S. = Column S.S./$m - 1$ | $F_c$ = C.M.S./ M.S.E. | |
| Treatment | $m - 1$ | T.S.S. | T.M.S. = T.S.S./$t - 1$ | $F_t$ = T.M.S./ M.S.E. | |
| Error | $(m - 1)$ $(m - 2)$ | E.S.S. | M.S.E. = E.S.S./$(m - 1)$ $(m - 2)$ | | |
| Total (n) | $m^2 - 1$ | Total S.S. | | | |

$$SEm = \sqrt{MS_{E}/_{r}}.$$

CD at 5% if treatment $F$ is significant $= SEm \times \sqrt{2} \times t_{0.05,ne}$,

$$CV\% = \frac{\sqrt{MS_E}}{\overline{Y}} \times 100.$$

## LSD Using R

### ***Required Package: "doebioresearch"***

*Required data format*

The data should be stored in a .csv (comma-separated values) file format, arranged in the following order:

A1 | fx | Row

| | A | B | C | D | E | F |
|---|---|---|---|---|---|---|
| 1 | Row | Column | Treatmer | Yield | Plant_Height | |
| 2 | 1 | 1 | D | 39 | 12 | |
| 3 | 1 | 2 | A | 24.1 | 15 | |
| 4 | 1 | 3 | E | 26.1 | 12 | |
| 5 | 1 | 4 | B | 37 | 14 | |
| 6 | 1 | 5 | C | 42 | 18 | |
| 7 | 2 | 1 | E | 21 | 19 | |
| 8 | 2 | 2 | B | 38 | 20 | |
| 9 | 2 | 3 | A | 24 | 22 | |
| 10 | 2 | 4 | C | 39 | 13 | |
| 11 | 2 | 5 | D | 33.1 | 14 | |
| 12 | 3 | 1 | C | 35.6 | 16 | |
| 13 | 3 | 2 | E | 33.5 | 19 | |
| 14 | 3 | 3 | B | 38.1 | 12 | |
| 15 | 3 | 4 | D | 40 | 21 | |
| 16 | 3 | 5 | A | 24 | 23 | |
| 17 | 4 | 1 | A | 30 | 24 | |
| 18 | 4 | 2 | C | 31 | 25 | |
| 19 | 4 | 3 | D | 46 | 26 | |
| 20 | 4 | 4 | E | 28 | 24 | |
| 21 | 4 | 5 | B | 44.9 | 25 | |
| 22 | 5 | 1 | B | 44 | 22 | |
| 23 | 5 | 2 | D | 29 | 21 | |
| 24 | 5 | 3 | C | 41 | 13 | |
| 25 | 5 | 4 | A | 26 | 15 | |
| 26 | 5 | 5 | E | 24 | 14 | |
| 27 | | | | | | |
| 28 | | | | | | |

Data must be arranged in row and column format with name of treatments

Read this data file in RStudio and launch the package "doebioresearch"

### *Code for Latin Square Analysis*

Code: lsd_ana=lsd(data[4],data$Treatment,data$Row,data$Column,1)

Write this code in script editor and run

```
RStudio
File Edit Code View Plots Session Build Debug Profile Tools Help
Go to file/function  Addins
Untitled1*  j_an
Source on Save
setwd("D:/mydata")
data=read.csv("LSD_data.csv")

lsd_ana=lsd(data[4],data$Treatment,data$Row,data$Column,1)
lsd_ana
```

**Description of code**

- **data:** dependent variables
- **treatmentvector:** vector having different treatments
- **row:** rows in the experimental design
- **column:** columns in the experimental design
- **multipleComparisonTest:** A parameter to specify the type of multiple comparison test to be used. The options are '0' for no test, '1' for the Least Significant Difference (LSD) test, '2' for the Duncan test, and '3' for the Tukey's Honestly Significant Difference (HSD) test.

**Result in console**

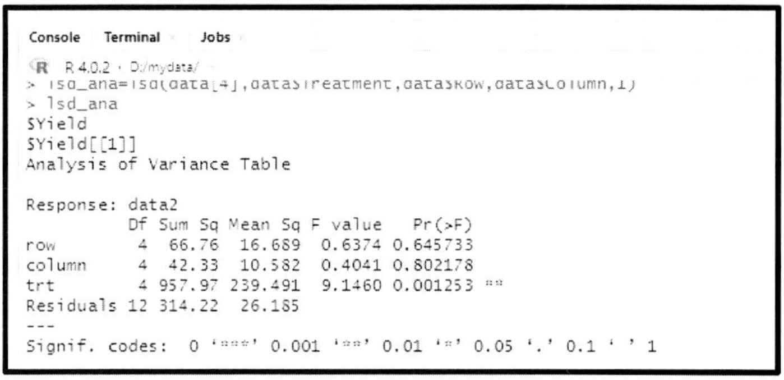

```
Console  Terminal  Jobs
R 4.0.2 · D:/mydata/
> lsd_ana=lsd(data[4],data$Treatment,data$Row,data$Column,1)
> lsd_ana
$Yield
$Yield[[1]]
Analysis of Variance Table

Response: data2
          Df Sum Sq Mean Sq F value   Pr(>F)
row        4  66.76  16.689  0.6374 0.645733
column     4  42.33  10.582  0.4041 0.802178
trt        4 957.97 239.491  9.1460 0.001253 **
Residuals 12 314.22  26.185
---
Signif. codes:  0 '***' 0.001 '**' 0.01 '*' 0.05 '.' 0.1 ' ' 1
```

## Complete result from text file

```
$Yield
$Yield[[1]]
Analysis of Variance Table

Response: data2
          Df Sum Sq Mean Sq F value   Pr(>F)
row        4  66.76  16.689  0.6374 0.645733
column     4  42.33  10.582  0.4041 0.802178
trt        4 957.97 239.491  9.1460 0.001253 **
Residuals 12 314.22  26.185
---
Signif. codes:  0 '***' 0.001 '**' 0.01 '*' 0.05 '.' 0.1 ' ' 1

$Yield[[2]]
[1] "R Square 0.773"

$Yield[[3]]

        Shapiro-Wilk normality test

data:  model$residuals
W = 0.98241, p-value = 0.9283


$Yield[[4]]
[1] "Normality assumption is not violated"

$Yield[[5]]
[1] "SEm 2.2885 , SEd 3.2364"

$Yield[[6]]
```

```
[1] "The treatment means of one or more treatments are not same, so go for multiple comparison test"

$Yield[[7]]
$Yield[[7]][[1]]
 MSerror Df  Mean      CV  t.value     LSD
 26.1854 12 33.536 15.25873 2.178813 7.051468

$Yield[[7]][[2]]
 data2 groups
B 40.40      a
C 37.72      a
D 37.42      a
E 26.52      b
A 25.62      b
```

# 11

# Split and Strip Plot Designs

This chapter focuses on split plot and strip plot designs, essential tools in experimental design known for effectively handling complex experiments. It covers the key principles and applications of these designs, emphasizing how they can be used to address specific research challenges. The chapter also provides practical guidance on conducting analyses of split and strip plot designs using R software, making it a valuable resource for researchers seeking to apply these methods in their work.

## Split Plot Design

The split-plot design is a statistical approach widely used in experiments where specific conditions apply:

- Hierarchical arrangement of treatments: This design is ideal for experiments involving two levels of experimental units, typically "whole plots" and "subplots." The main treatments are applied to the whole plots, while the sub-treatments are applied within these to the subplots.
- Adherence to experimental design principles: The split-plot design follows the principles of randomization, replication, and local control, but with a hierarchical twist. The main treatments are randomized across the whole plots, and the sub-treatments are randomized within these main plots.
- Resource management in unequal experimental conditions: This design is particularly useful when some factors require larger experimental units or when changing levels of one factor is more resource-intensive than changing levels of another factor.

## Strip Plot Design

The strip plot design, also known as the split-block design, is another experimental design framework used under certain experimental conditions:

- Two-dimensional variation control: The strip plot design is effective when experiments need to control for two factors, similar to the Latin square design. However, it provides more flexibility in the number of levels for each factor.
- Principles of randomization and local control: In this design, one factor is applied in strips across the field in one direction (e.g., horizontally) and the second factor is applied perpendicularly (e.g., vertically). This creates a matrix-like structure, allowing for two-factor interaction analysis.
- Resource optimization for two-factor experiments: The strip plot design efficiently uses resources by allowing for two factors to be tested simultaneously, especially suited for a two-factor experiment in which the desired precision for measuring the interaction effect between the two factors is higher than that for measuring the main effect of either one of the two factors.

## Layout and Randomization in Split Plot Design

In the split-plot design, the allocation of main plot treatments and sub-plot treatments is determined by the desired level of precision. The factor requiring greater precision is assigned to the sub-plots, while the main plot factor is selected accordingly.

Next, the replication is divided into several main plots, corresponding to the main plot treatments. Each main plot is further subdivided into sub-plots, depending on the number of sub-plot treatments. The allocation of main plot treatments is randomized, similar to the process in a randomized block design (RBD).

Within each main plot, the sub-plot treatments are also allocated randomly, following a process similar to RBD. This two-stage randomization ensures unbiased treatment assignment. This procedure is independently applied to all replications, maintaining consistency and reliability across the experiment.

## Layout and Randomization in Strip Plot Design

The process of randomization and layout in the Strip Plot Design involves two distinct steps:

### *Step 1: Horizontal plot assignment*

Begin by dividing the experimental area into "r" blocks, and within each block, further divide it into "a" horizontal strips. Utilize a randomization procedure similar to that of randomized block design (RBD). Independently and separately, assign the levels of the first factor to the horizontal strips in each of the "r" blocks.

### *Step 2: Vertical plot assignment*

Within each block, further divide it into "b" vertical strips. Apply the randomization procedure as if conducting an RBD with "b" treatments and "r" replications. Randomly allocate the "b" levels to the "b" vertical strips in each block, independently and separately.

This two-step process ensures a randomized assignment of treatments to both horizontal and vertical strips, maintaining the principles of randomization and independence throughout the experiment.

## ANOVA for Split Plot Design

The analysis of variance in a Split Plot Design comprises two distinct parts, each corresponding to the main plots and subplots. Here's how it's done:

### *For Main Plot Analysis*

I. Create a replication X main plot treatments table, forming a two-way table.

II. Compute the sum of squares for replication, main plot treatments, and error (a).

### *For Sub-Plot Analysis*

I. Form a main plot X sub-plot treatments table, establishing another two-way table.

II. Calculate the sums of squares for sub-plot treatments and the interaction between main plot and sub-plot treatments.

III. Determine the error (b) sum of squares using the residual method.

The analysis of the variance table for a split-plot design with "m" main plot treatments and "s" sub-plot treatments is as follows:

ANOVA for split plot with factor A (m levels in main plots) and Factor B (s levels in sub-plots):

| SV | *d.f.* | SS | MS | *F* |
|---|---|---|---|---|
| Replication | $r - 1$ | RSS | RMS | RMS/EMS (a) |
| A | $m - 1$ | ASS | AMS | AMS/EMS (a) |
| Error (a) | $(r - 1)(m - 1)$ | ESS (a) | EMS (a) | |
| B | $s - 1$ | BSS | BMS | BMS/EMS (b) |
| AB | $(m - 1)(s - 1)$ | ABSS | ABMS | ABMS/EMS (b) |
| Error (b) | $m(r - 1)(s - 1)$ | ESS (b) | EMS (b) | |
| v | $rms - 1$ | TSS | | |

## ANOVA for Strip Plot Design

| Source | *D.F.* | *S* | M.S. | Cal *F* |
|---|---|---|---|---|
| Repli. | $(r - 1)$ | SSR | MS(R) | MSR/MSE(a) |
| Factor A | $(a - 1)$ | SSA | MS(A) | MSA/MSE(a) |
| Error (a) | $(r - 1)*(a - 1)$ | SS (R) – SS(A) | MSE(a) | |
| Factor B | $(b - 1)$ | SSB | MS(B) | MSB/MSE(b) |
| Error (b) | $(r - 1)*(b - 1)$ | SS(R) – SS(B) | MSE(b) | |
| A × B | $(a - 1)*(b - 1)$ | SS(A) – SS(B) | MS(AB) | MS(A × B)/MSE(c) |
| Error (c) | $(r - 1)*(a - 1)*(b - 1)$ | By difference | MSE(c) | |
| Total | rab – 1 | $\Sigma Y^2_{ijk}$ – C.F. | | |

## Split Plot Design in R

For a design with two factors, the arrangement of the input data file should follow the format outlined below:

| | A | B | C | D | E |
|---|---|---|---|---|---|
| 1 | Replicati | Date_of_Sowing | Varities | Yield | Plant_Height |
| 2 | 1 | 1 | 1 | 4800 | 98 |
| 3 | 1 | 1 | 2 | 4600 | 104 |
| 4 | 1 | 1 | 3 | 5000 | 106 |
| 5 | 1 | 1 | 4 | 5120 | 110 |
| 6 | 1 | 1 | 5 | 5130 | 120 |
| 7 | 1 | 1 | 6 | 5140 | 125 |
| 8 | 1 | 2 | 1 | 4700 | 98 |
| 9 | 1 | 2 | 2 | 4969 | 96 |
| 10 | 1 | 2 | 3 | 5013 | 99 |
| 11 | 1 | 2 | 4 | 4890 | 100 |
| 12 | 1 | 2 | 5 | 5150 | 120 |
| 13 | 1 | 2 | 6 | 5070 | 126 |
| 14 | 2 | 1 | 1 | 4830 | 120 |
| 15 | 2 | 1 | 2 | 5110 | 123 |
| 16 | 2 | 1 | 3 | 4920 | 126 |
| 17 | 2 | 1 | 4 | 4900 | 123 |
| 18 | 2 | 1 | 5 | 4880 | 150 |
| 19 | 2 | 1 | 6 | 4986 | 140 |
| 20 | 2 | 2 | 1 | 5112 | 130 |
| 21 | 2 | 2 | 2 | 5110 | 135 |
| 22 | 2 | 2 | 3 | 5230 | 102 |
| 23 | 2 | 2 | 4 | 5120 | 123 |
| 24 | 2 | 2 | 5 | 5215 | 132 |
| 25 | 2 | 2 | 6 | 5200 | 125 |
| 26 | | | | | |

**To perform a split plot design analysis with two factors, date of sowing and varieties, you can use the following code:**

## Code for the split-plot design

split_ana=splitplot(data[4],data$Replication,data$Date_of_Sowing,data$Varities,1)

Copy and paste this code into the script editor in RStudio and then run the code to perform the split plot analysis for the specified factors.

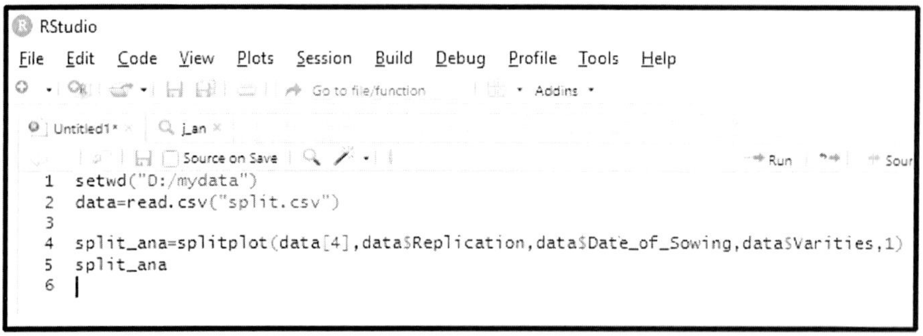

RStudio

File Edit Code View Plots Session Build Debug Profile Tools Help

Untitled1* i_an

Source on Save Run

```
setwd("D:/mydata")
data=read.csv("split.csv")

split_ana=splitplot(data[4],data$Replication,data$Date_of_Sowing,data$Varities,1)
split_ana
```

## Code arguments overview

- **data:** Represents the dependent variables in the dataset.
- **block:** A vector containing information about replications or blocks in the experiment.
- **main.plot:** A vector containing the levels or treatments of the main plots.
- **sub.plot:** A vector containing the levels or treatments of the sub-plots.
- **mean.comparison.test:** An option to specify the type of mean comparison test to be performed. Choices include "0" for no test, "1" for the least significant difference (LSD) test, "2" for the Duncan test, and "3" for the Tukey's honestly significant difference (HSD) test.

The input data should be structured in a format that includes the following essential information:

- **Yield data:** This section should contain the actual yield measurements for each plot or experimental unit.
- **Net plot size:** Information about the size of the net plot, which serves as the basis for conversion to a hectare (ha) basis.

Ensure that the data is organized and labeled appropriately to facilitate the conversion process.

**Input data format**

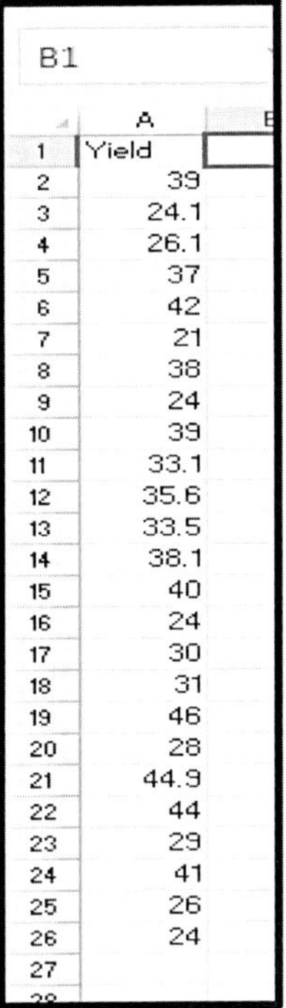

| | A |
|---|---|
| 1 | Yield |
| 2 | 39 |
| 3 | 24.1 |
| 4 | 26.1 |
| 5 | 37 |
| 6 | 42 |
| 7 | 21 |
| 8 | 38 |
| 9 | 24 |
| 10 | 39 |
| 11 | 33.1 |
| 12 | 35.6 |
| 13 | 33.5 |
| 14 | 38.1 |
| 15 | 40 |
| 16 | 24 |
| 17 | 30 |
| 18 | 31 |
| 19 | 46 |
| 20 | 28 |
| 21 | 44.9 |
| 22 | 44 |
| 23 | 29 |
| 24 | 41 |
| 25 | 26 |
| 26 | 24 |
| 27 | |

**To convert the yield data from kg/plot to the desired format, you can use the following code:**

```
Yield_conv=yieldconvert(data$Yield,10,5,1)
```

Write this code in the script editor and run this code for yield conversion in desire format

RStudio

File Edit Code View Plots Session Build Debug Profile Tools Help

Untitled1* j_an

Source on Save

```
setwd("D:/mydata")
data=read.csv("yield.csv")

Yield_conv=yieldconvert(data$Yield,10,5,1)
Yield_conv
```

## Description of code

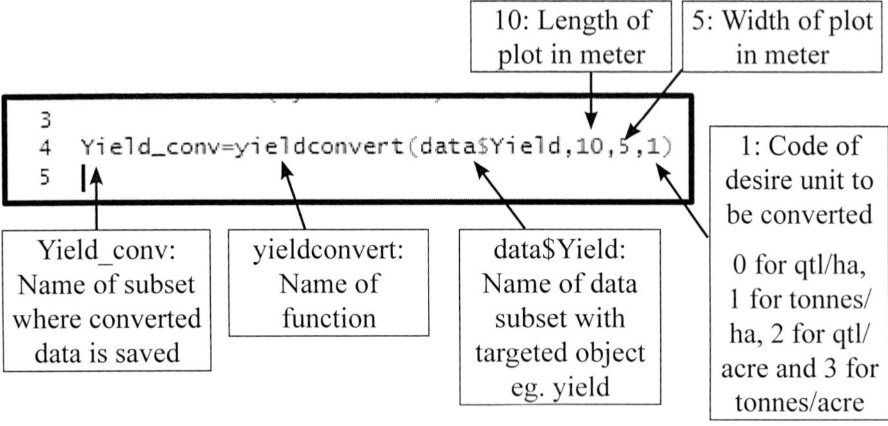


## Output in console

```
>
> Yield_conv
[[1]]
[1] "Converted to Tonnes/Ha"

[[2]]
 [1] 7.80 4.82 5.22 7.40 8.40 4.20 7.60 4.80 7.80 6.62 7.12 6.70 7.62 8.00 4.80 6.00 6.20
[18] 9.20 5.60 8.98 8.80 5.80 8.20 5.20 4.80

>
```

## Strip Plot Design in R

### *SPD: Strip Plot Design (doebioresearch)*

Required data format

Data should be saved in .csv file format in below order

This design has two factors. So, the arrangement of the input data file should be like below

| | A | B | C | D | E |
|---|---|---|---|---|---|
| 1 | Replicatic | Irrigation | Fertilizer | Yield | Height |
| 2 | 1 | 1 | 1 | 4800 | 98 |
| 3 | 1 | 1 | 2 | 4600 | 104 |
| 4 | 1 | 1 | 3 | 5000 | 106 |
| 5 | 1 | 1 | 4 | 5120 | 110 |
| 6 | 1 | 1 | 5 | 5130 | 120 |
| 7 | 1 | 1 | 6 | 5140 | 125 |
| 8 | 1 | 2 | 1 | 4700 | 98 |
| 9 | 1 | 2 | 2 | 4969 | 96 |
| 10 | 1 | 2 | 3 | 5013 | 99 |
| 11 | 1 | 2 | 4 | 4890 | 100 |
| 12 | 1 | 2 | 5 | 5150 | 120 |
| 13 | 1 | 2 | 6 | 5070 | 126 |
| 14 | 2 | 1 | 1 | 4830 | 120 |
| 15 | 2 | 1 | 2 | 5110 | 123 |
| 16 | 2 | 1 | 3 | 4920 | 126 |
| 17 | 2 | 1 | 4 | 4900 | 123 |
| 18 | 2 | 1 | 5 | 4880 | 150 |
| 19 | 2 | 1 | 6 | 4986 | 140 |

In the strip plot design, each block or replication is divided into a grid of vertical and horizontal strips and the size of each strip corresponds to the levels of the respective factors being studied. These vertical and horizontal strips are always arranged perpendicular to each other in the field. The layout in the field resembles a grid pattern where treatments are assigned to specific intersections of vertical and horizontal strips, allowing for the study of multiple factors simultaneously.

| **Replication 1** | | | | | | |
|---|---|---|---|---|---|---|
| | Irrigation 1 | Irrigation 2 | Irrigation 3 | Irrigation 4 | Irrigation 5 | Irrigation 6 |
| Fertilizer 1 | | | | | | |
| Fertilizer 2 | | | | | | |
| **Replication 2** | | | | | | |
| | Irrigation 6 | Irrigation 5 | Irrigation 4 | Irrigation 3 | Irrigation 2 | Irrigation 1 |
| Fertilizer 2 | | | | | | |
| Fertilizer 1 | | | | | | |
| **Replication 3** | | | | | | |
| | Irrigation 5 | Irrigation 1 | Irrigation 2 | Irrigation 6 | Irrigation 3 | Irrigation 4 |
| Fertilizer 1 | | | | | | |
| Fertilizer 2 | | | | | | |

## Code for the split-plot design

**Code:**

```
SPD_ana=stripplot(data[4:5],data$Replication,data$Irrigation,data$Fertilizer,1)
```

Write this code in the script editor and run the code

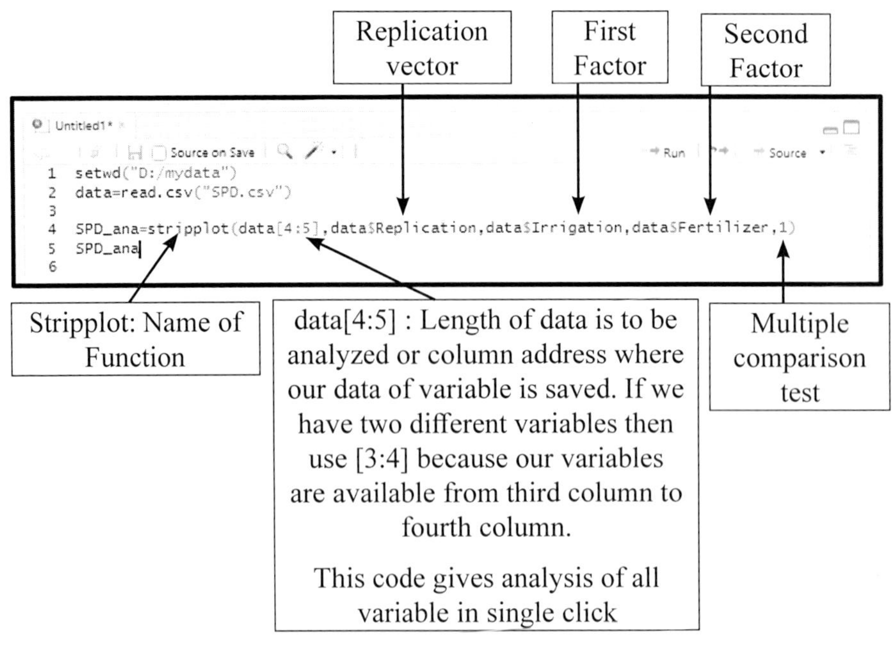

Arguments

- **Data:** Dependent variables
- **Block:** Vector containing replications
- **Column:** Vector containing column strip levels
- **Row:** Vector containing row strip levels
- **Mean comparison test:** 0 for no test, 1 for LSD test, 2 for Duncan test, and 3 for HSD test

**Output in console**

```
Console   Terminal   Jobs
R 4.0.2 · D:/mydata/
> setwd("D:/mydata")
> data=read.csv("SPD.csv")
> SPD_ana=stripplot(data[4:5],data$Replication,data$Irrigation,data$Fertilizer,1)
> SPD_ana
$Yield
$Yield[[1]]
$Yield[[1]][[1]]
            Df       Sum Sq      Mean Sq   F value     Pr(>F)
block        2   61557.0556   30778.5278 0.8915086 0.4402683
Column       1     186.7778     186.7778 0.0010000 0.9776000
Ea           2  328175.7222  164087.8611        NA        NA
Row          5  510277.2222  102055.4444 3.0280000 0.0640000
Eb          10  337066.2778   33706.6278        NA        NA
Interaction  5  125446.5556   25089.3111 0.7267189 0.6190384
Ec          10  345240.9444   34524.0944        NA        NA

$Yield[[1]][[2]]
[1] "CV(a): 8.063 , CV(b) : 3.655 , CV(c) : 3.699"

$Yield[[1]][[3]]
[1] "R Square 0.798"

$Yield[[1]][[4]]

        Shapiro-Wilk normality test

data:  model$residuals
W = 0.98651, p-value = 0.931
```

To obtain the lengthy output of the code with various statistical values, you can save it in a .txt file using the "sink" function, similar to how it's done in CDR analysis.

## Complete output in text file

$Yield

$Yield[[1]]

$Yield[[1]][[1]]

| | Df | Sum Sq | Mean Sq | F value | Pr(>F) |
|---|---|---|---|---|---|
| block | 2 | 61557.0556 | 30778.5278 | 0.8915086 | 0.4402683 |
| Column | 1 | 186.7778 | 186.7778 | 0.0010000 | 0.9776000 |
| Ea | 2 | 328175.7222 | 164087.8611 | NA | NA |
| Row | 5 | 510277.2222 | 102055.4444 | 3.0280000 | 0.0640000 |
| Eb | 10 | 337066.2778 | 33706.6278 | NA | NA |
| Interaction | 5 | 125446.5556 | 25089.3111 | 0.7267189 | 0.6190384 |
| Ec | 10 | 345240.9444 | 34524.0944 | NA | NA |

$Yield[[1]][[2]]

[1] "CV(a): 8.063 , CV(b) : 3.655 , CV(c) : 3.699"

$Yield[[1]][[3]]

[1] "R Square 0.798"

$Yield[[1]][[4]]

Shapiro-Wilk normality test

data: model$residuals

W = 0.98651, p-value = 0.931

$Yield[[1]][[5]]

[1] "Normality assumption is not violated"

$Yield[[1]][[6]]

```
[1] "All the column factor level means are same so dont go for any multiple comparison test"

$Yield[[1]][[7]]
$Yield[[1]][[7]][[1]]
  MSerror Df     Mean       CV  t.value      LSD
 164087.9  2 5023.611 8.063474 4.302653 580.9694

$Yield[[1]][[7]][[2]]
  dependent.var groups
1      5025.889      a
2      5021.333      a


$Yield[[1]][[8]]
[1] "All the row factor factor level means are same so dont go for any multiple comparison test"

$Yield[[1]][[9]]
$Yield[[1]][[9]][[1]]
  MSerror Df     Mean       CV  t.value      LSD
 33706.63 10 5023.611 3.654615 2.228139 236.1779

$Yield[[1]][[9]][[2]]
  dependent.var groups
6      5154.333      a
5      5075.833      a
4      5070.000      a
3      5053.833      a
2      5013.167      a
1      4774.500      b
```

```
$Yield[[1]][[10]]
[1] "All the interaction level means are same so dont go for any multiple comparison test"


$Yield[[1]][[11]]
$Yield[[1]][[11]][[1]]
  MSerror Df     Mean       CV  t.value     LSD
 34524.09 10 5023.611 3.698666 2.228139 338.032


$Yield[[1]][[11]][[2]]
    dependent.var groups
2:6      5173.333      a
1:6      5135.333      a
2:5      5095.000      a
2:3      5074.333      a
1:4      5070.000      a
2:4      5070.000      a
2:2      5069.667      a
1:5      5056.667      a
1:3      5033.333      a
1:2      4956.667     ab
1:1      4903.333     ab
2:1      4645.667      b
```

# 12

# Visualization of Output

In this chapter, we will explore multiple comparison tests, with a particular emphasis on three significant tests: least significant difference (LSD), Bonferroni, and the techniques to visualize their outcomes. We aim to facilitate a comprehensive understanding of how to conduct and interpret the results of these comparison tests visualization using the R software.

## Treatment Comparison

The "doebioresearch" package is commonly used for testing the significance of treatment groups in various experimental designs such as randomized block design (RBD), completely randomized design (CRD), and least significant difference (LSD). This package excels in conducting multiple comparison tests like LSD, Duncan's multiple range test (DMRT), and Tukey's honestly significant difference (HSD). It efficiently calculates all comparisons in the background, resulting in the generation of only group labels, as previously demonstrated in the output of the "doebioresearch" package. Additionally, the "agricolae" package is another useful tool that provides detailed values for multiple comparison tests. These three tests—LSD, DMRT, and HSD—are particularly popular in agricultural experiments and form the focus of our discussion.

I. **Least significant difference (LSD) test**

- To conduct the least significant difference test for comparing treatment means, the "agricolae" package is essential.
- To install the "agricolae" package, use the following code:
- install.packages ("agricolae")
- After installation, activate the package with this command:
- library ("agricolae")
- The data frame required for analysis in this context remains consistent with what is needed for randomized block design (RBD), completely

randomized design (CRD), LSD, and variability analysis, among others. As an illustration, consider an example involving variability analysis.

An overview of the input data file is provided, similar to what one might see in an Excel spreadsheet.

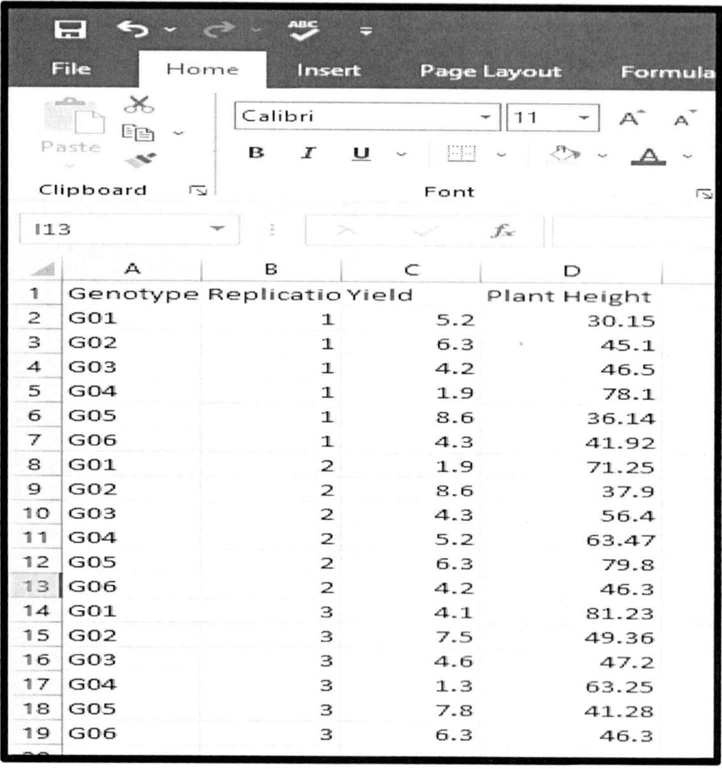

| | A | B | C | D |
|---|---|---|---|---|
| 1 | Genotype | Replicatio | Yield | Plant Height |
| 2 | G01 | 1 | 5.2 | 30.15 |
| 3 | G02 | 1 | 6.3 | 45.1 |
| 4 | G03 | 1 | 4.2 | 46.5 |
| 5 | G04 | 1 | 1.9 | 78.1 |
| 6 | G05 | 1 | 8.6 | 36.14 |
| 7 | G06 | 1 | 4.3 | 41.92 |
| 8 | G01 | 2 | 1.9 | 71.25 |
| 9 | G02 | 2 | 8.6 | 37.9 |
| 10 | G03 | 2 | 4.3 | 56.4 |
| 11 | G04 | 2 | 5.2 | 63.47 |
| 12 | G05 | 2 | 6.3 | 79.8 |
| 13 | G06 | 2 | 4.2 | 46.3 |
| 14 | G01 | 3 | 4.1 | 81.23 |
| 15 | G02 | 3 | 7.5 | 49.36 |
| 16 | G03 | 3 | 4.6 | 47.2 |
| 17 | G04 | 3 | 1.3 | 63.25 |
| 18 | G05 | 3 | 7.8 | 41.28 |
| 19 | G06 | 3 | 6.3 | 46.3 |

To create a data frame for variability analysis as outlined in Chapter 7, save it in a .csv format, read the appropriate subset, and verify it in the console.

**Script editor view**

```
RStudio
File Edit Code View Plots Session Build Debug
Go to file/function
HSD test.R   Mean Comparison with graphs.R*   Untitl
Source on Save
1 setwd("D:/01 R book/Data")
2 Analysis=read.csv("variability.csv")
3 data
```

**Output in console**

```
Console   Terminal   Background Jobs
R 4.0.2 · D:/01 R book/Data/
> Analysis=read.csv("variability.csv")
> Analysis
   Genotype Replication Yield Plant.Height
1       G01           1   5.2        30.15
2       G02           1   6.3        45.10
3       G03           1   4.2        46.50
4       G04           1   1.9        78.10
5       G05           1   8.6        36.14
6       G06           1   4.3        41.92
7       G01           2   1.9        71.25
8       G02           2   8.6        37.90
9       G03           2   4.3        56.40
10      G04           2   5.2        63.47
11      G05           2   6.3        79.80
12      G06           2   4.2        46.30
13      G01           3   4.1        81.23
14      G02           3   7.5        49.36
15      G03           3   4.6        47.20
16      G04           3   1.3        63.25
17      G05           3   7.8        41.28
18      G06           3   6.3        46.30
>
```

This data contains six genotypes/treatments with three replications for two variables.

***Step 1: Perform ANOVA and SAVE it in a separate subset like below:***

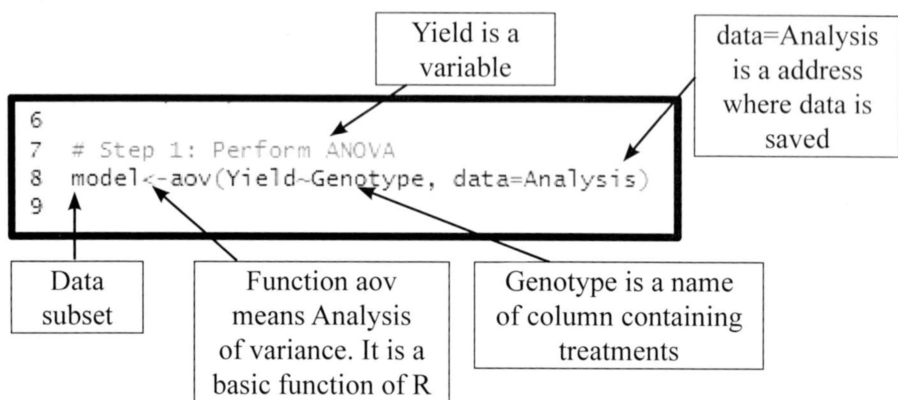


After executing the script, the results will be stored in the specified subset, such as "model." The LSD test, which is a function within the "agricolae" package, can then be performed. To do this, first, ensure that the "agricolae" package is loaded and then proceed to write and run the code for the LSD test in the script editor.

Code: out=LSD.test(model,"Genotypes", p.adj="bonferroni")

**Description of function and code**

- **out:** Subset where analyzed data will be saved.
- **LSD.test:** This function, part of the "agricolae" package, performs the LSD test on results contained within the "model" subset (Note: "model" contains the results of the analysis of variance of the data).
- **"Genotype":** This refers to the column name in the data file that contains the treatment groups.
- **p.adj:** This parameter allows for the specification of the method used for adjusting means. Options include "holm," "hommel," "hochberg," "bonferroni," "BH," "BY," "fdr," or "none" if no adjustment is needed.
- **Other extra arguments** of this function
  - **group=TRUE:** This option outputs the mean value along with the groups.
  - **main=NULL:** Use this to add a title to your study, if desired.

## Write above code in script editor and check the output in console

```
# step 2: Perform LSD test
library("agricolae")
out=LSD.test(model,"Genotype", p.adj="bonferroni")
out
```

## Output in console

```
Console   Terminal   Background Jobs
R 4.0.2 · D:/01 R book/Data/
> View(out)
> out
$statistics
   MSerror Df     Mean       CV  t.value      MSD
  1.894444 12 5.144444 26.75485 3.648889 4.100681

$parameters
        test  p.ajusted   name.t ntr alpha
  Fisher-LSD bonferroni Genotype   6  0.05

$means
       Yield       std r      LCL      UCL Min Max  Q25 Q50  Q75
G01 3.733333 1.6802778 3 2.001922 5.464745 1.9 5.2 3.00 4.1 4.65
G02 7.466667 1.1503623 3 5.735255 9.198078 6.3 8.6 6.90 7.5 8.05
G03 4.366667 0.2081666 3 2.635255 6.098078 4.2 4.6 4.25 4.3 4.45
G04 2.800000 2.1000000 3 1.068589 4.531411 1.3 5.2 1.60 1.9 3.55
G05 7.566667 1.1676187 3 5.835255 9.298078 6.3 8.6 7.05 7.8 8.20
G06 4.933333 1.1846237 3 3.201922 6.664745 4.2 6.3 4.25 4.3 5.30

$comparison
NULL

$groups
       Yield groups
G05 7.566667      a
G02 7.466667      a
G06 4.933333     ab
G03 4.366667     ab
G01 3.733333     ab
G04 2.800000      b
```

The output contains extensive information and we can print it using the "sink" function as previously described.

Additionally, we can perform honestly significant difference (HSD) and Duncan's multiple range test (DNMRT) for further analysis.

## Visualization of Test of Comparison of Treatment

To visualize the comparison of treatments, we aim to create a bar graph that includes standard deviation and treatment group information derived from the results of the multiple comparison test, specifically LSD.tst.

Before constructing the bar graph, we need certain data, which we have obtained from the previous analysis, specifically mean values, standard deviations, and treatment groups.

In the next step, we create a new subset based on these groups for visualization.

### *Step 2: Create new subset groups*

```
# step 3: Creat new subset containing groups
groups=out$groups
groups
```

**Run this code and check in the console**

```
Console    Terminal    Background Jobs
R 4.0.2 · D:/01 R book/Data/
> groups
        Yield groups
G05 7.566667      a
G02 7.466667      a
G06 4.933333     ab
G03 4.366667     ab
G01 3.733333     ab
G04 2.800000      b
>
```

In this scenario, various groups are delineated by alphabets, with genotypes or treatments sorted in descending order based on the yield variable. To create a bar graph in R using these groups, you need to extract them into a separate subset. This can be achieved by using the following code in the script editor, which will create a new subset named "labels."

```
labels=groups$groups
labels
```

**See in the console**

```
> labels=groups$groups
> labels
[1] "a"  "a"  "ab" "ab" "ab" "b"
>
```

Now we need mean and standard deviations of genotypes or treatments. So, write code in the script editor and run.

```
# step 4: Creat new subset containing means
mean=out$means
mean
```

**Check in the console**

```
> # step 4: Creat new subset containing means
> mean=out$means
> mean
       Yield       std r      LCL      UCL Min Max  Q25 Q50  Q75
G01 3.733333 1.6802778 3 2.001922 5.464745 1.9 5.2 3.00 4.1 4.65
G02 7.466667 1.1503623 3 5.735255 9.198078 6.3 8.6 6.90 7.5 8.05
G03 4.366667 0.2081666 3 2.635255 6.098078 4.2 4.6 4.25 4.3 4.45
G04 2.800000 2.1000000 3 1.068589 4.531411 1.3 5.2 1.60 1.9 3.55
G05 7.566667 1.1676187 3 5.835255 9.298078 6.3 8.6 7.05 7.8 8.20
G06 4.933333 1.1846237 3 3.201922 6.664745 4.2 6.3 4.25 4.3 5.30
>
```

In this output, since the treatment names lack a header, we can add one using the "tibble" package. First, download and load the package in R. Then, use the provided code to insert a header into the mean subset.

**Write the below code to add a header in the mean subset**

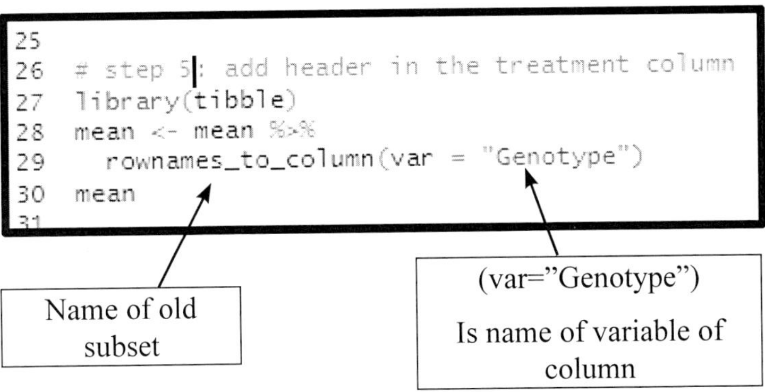


In R, '%>%', known as the **forward pipe operator**, facilitates the chaining of commands, allowing values or the results of expressions to be passed into subsequent function calls or expressions. The "**rownames_to_column**" function, provided by the "*tibble*" package, is used to add a header to the first column.

**Check in the console**

```
> mean
  Genotype    Yield       std r      LCL      UCL Min Max  Q25 Q50  Q75
1      G01 3.733333 1.6802778 3 2.001922 5.464745 1.9 5.2 3.00 4.1 4.65
2      G02 7.466667 1.1503623 3 5.735255 9.198078 6.3 8.6 6.90 7.5 8.05
3      G03 4.366667 0.2081666 3 2.635255 6.098078 4.2 4.6 4.25 4.3 4.45
4      G04 2.800000 2.1000000 3 1.068589 4.531411 1.3 5.2 1.60 1.9 3.55
5      G05 7.566667 1.1676187 3 5.835255 9.298078 6.3 8.6 7.05 7.8 8.20
6      G06 4.933333 1.1846237 3 3.201922 6.664745 4.2 6.3 4.25 4.3 5.30
>
```

When you check the console, you'll notice that the column is now named "treatment." However, these treatments are ordered by their names, whereas the groups are arranged in ascending order based on the value of the variable (in this case, Yield). To address this, you can write and execute the following code in the script editor.

**Write below code in script editor and run this code**

```

# step 6: Arrange treatments in descending order according to variable
arrangedtreatments=arrange(mean, desc(Yield))
arrangedtreatments
```

**Description of code**

- **arrangedtreatments:** Name of subset where arranged treatments will be saved
- **arrange:** Name of function provided by basic packages of R
- **mean:** Subset, which we want to arrange as per our requirement
- **desc:** Function to arrange data in descending order according to variable
- **(Yield):** Variable name, the data will be arranged in descending order concerning the "Yield" variable.

**Check-in console**

```
> arrangedtreatments
  Genotype    Yield       std r      LCL      UCL Min Max  Q25 Q50  Q75
1      G05 7.566667 1.1676187 3 5.835255 9.298078 6.3 8.6 7.05 7.8 8.20
2      G02 7.466667 1.1503623 3 5.735255 9.198078 6.3 8.6 6.90 7.5 8.05
3      G06 4.933333 1.1846237 3 3.201922 6.664745 4.2 6.3 4.25 4.3 5.30
4      G03 4.366667 0.2081666 3 2.635255 6.098078 4.2 4.6 4.25 4.3 4.45
5      G01 3.733333 1.6802778 3 2.001922 5.464745 1.9 5.2 3.00 4.1 4.65
6      G04 2.800000 2.1000000 3 1.068589 4.531411 1.3 5.2 1.60 1.9 3.55
>
```

Now, you should compare the order of treatments between the "group" subset and the newly created "*arrangedtreatments*" subset. If the order of treatments in both subsets is similar, then you can proceed with developing a bar graph.

### *Step 3: Development of Bar graph*

To create a bar graph, the "ggplot2" package is necessary.

First, download and initialize this package.

Developing a plot that includes various pieces of information can be complex, often requiring multiple lines of code to incorporate all the desired elements at once. However, in this case, we will break down the process into separate codes. These individual codes can then be combined using the "+" sign to construct a comprehensive plot that includes all the required information.

**Code 1: Define data to develop a bar graph**

```
ggplot(data = arrangedtreatments, aes(Genotype, Yield))
```

**Description**

- **ggplot:** Name of function
- **data= arrangedtreatments:** name of data subset whose data will be used to develop a bar graph
- **aes()** specify the aesthetic mappings and their arguments should be the names of columns in the "arrangedtreatments" dataset. In this instance, you'll be quoting "Genotype" and "Yield" within "aes()" to define how these variables are displayed in the plot.

**Code 2: Define the length of the scale of the *Y*-axis. It should be more than the highest value of treatment as per the requirement**

```
scale_y_continuous(limits=c(0,12))
```

In this context, setting "limits=c(0, 12)" specifies that the *Y*-axis of the graph will start from zero and extend up to 12 units. This range should be determined based on the highest mean value plus the highest standard deviation in the data to adequately represent the entire range of values in the graph.

**Code 3: Develop bar graph**

```
geom_bar(stat = "identity," fill = "Blue")
```

**Description**

- **geom_bar():** Height of the bar proportional to the number of cases in each group (or if the weight aesthetic is supplied, the sum of the weights).
- **stat="identity":** Helps to take data from the provided subset without transforming it.
- **fill="Blue":** Used to fill color in bars.

**Code 4: Development of error bar**

```
geom_errorbar(aes(ymin=Yield-std, ymax=Yield+std), width=0.1)
```

- **aes:** Quoting function
- **ymin:** Minimum limit of error bar from mean
- **Yield:** Mean of variable available in subset ie. arrangedtreatments in this case
- **std:** Standard deviation of the data available in the subset
- **width:** Horizontal width of error bar

**Code 5: Addition of labels of mean comparison**

```
geom_text(aes(label = labels, y = Yield + std), angle=90, hjust=0.5, vjust = 0.3)
```

- **aes:** Quoting the labels on the bar graph
- **label:** Utilize the subset which has information on labels
- **y=Yield+std:** Defines the location of labels on the upper limit of the error bar
- **angle:** Defines the writing direction of the label. If it is 0, the label will written horizontally. If it is 90, the label will be written from the lower side to the upper side. If it is 180 it will write labels opposite
- **hjust:** Space between the label and upper limit of error bar
- **vjust:** Horizontal location of text on the upper limit of error bar

**Code 6: give a label to plot**

```
ggtitle ("Yield plot")
```

**Code 7: Define the position of the title**

```
theme (plot.title = element_text(hjust = 0.5))
```

## Code 8 Define the angle of labels on *x*-axis

theme (axis.text.x = element_text(angle = 0))

Now, merge all this code by placing one + sign between two codes as shown below in the script editor

### Script editor

```
# step 7: Developing bar graph
r=ggplot(data = arrangedtreatments, aes(Genotype, Yield)) +
  scale_y_continuous(limits=c(0,12)) +
  geom_bar(stat= "identity",fill= "Blue") +
  geom_errorbar(aes(ymin=Yield-std, ymax=Yield+std), width=0.2) +
  geom_text(aes(label = labels, y = Yield + std), angle=90, hjust=-0.5, vjust = 0.3) +
  ggtitle("Yield plot") +
  theme(plot.title = element_text(hjust = 0.5)) +
  theme(axis.text.x = element_text(angle = 0))
r
```

In the script editor, the entire "ggplot" script is stored in a separate subset named "r." After running the code, you should check both the console and the plot area. The console will display any messages or output from running the script, while the plot area will show the generated bar graph.

### Console

```
> r=ggplot(data = arrangedtreatments, aes(Genotype, Yield)) +
+   scale_y_continuous(limits=c(0,12)) +
+   geom_bar(stat= "identity",fill= "Blue") +
+   geom_errorbar(aes(ymin=Yield-std, ymax=Yield+std), width=0.2) +
+   geom_text(aes(label = labels, y = Yield + std), angle=90, hjust=-0.5, vjust = 0.3) +
+   ggtitle("Yield plot") +
+   theme(plot.title = element_text(hjust = 0.5)) +
+   theme(axis.text.x = element_text(angle = 0))
>
```

Console indicate code is run successfully. Now write "*r*" in the console and press enter

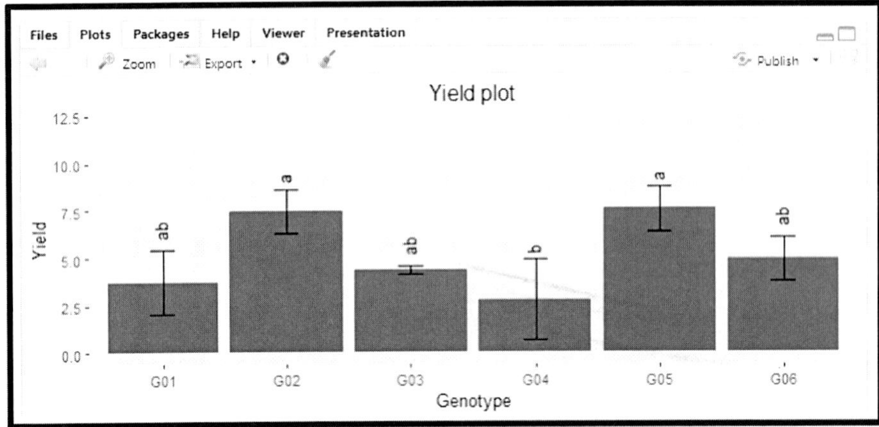


The bar graph currently displayed shows the yield variable for all the treatments considered. To create a similar bar graph for a different variable, such as plant height, you can follow the same sequence of code, with modifications only where necessary to accommodate the new variable.

For instance, in the script editor, simply replace instances of the yield variable with plant height in the relevant parts of the code. This will generate a bar graph for plant height, which can then be saved in a separate subset, such as "s," following the same method used for the yield variable graph.

## Script editor

```
# step 7: Developing bar graph
s=ggplot(data = arrangedtreatments, aes(Genotype, Plant.Height)) +
  scale_y_continuous(limits=c(0,95)) +
  geom_bar(stat= "identity",fill= "Red") +
  geom_errorbar(aes(ymin=Plant.Height-std, ymax=Plant.Height+std), width=0.2) +
  geom_text(aes(label = labels1, y = Plant.Height + std), angle=90, hjust=-0.5, vjust = 0.3) +
  ggtitle("Plant.Height plot") +
  theme(plot.title = element_text(hjust = 0.5)) +
  theme(axis.text.x = element_text(angle = 0))
s
```

### And see the results

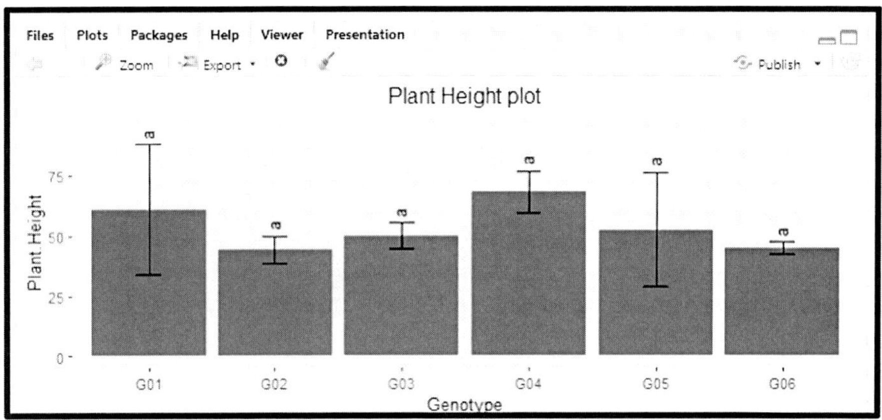


Now, you can combine the two different plots—one for Yield and the other for plant height—into a single frame using the "metan" package. Begin by initializing the "metan" package in R. Then, apply the appropriate code in the script editor to merge these plots into a unified display.

```
library(metan)

t=arrange_ggplot(r, s, ncol = 2)
t
```

The first plot, saved in the "r" subset, and the second plot, stored in the "s" subset, are intended to be combined side by side.

To achieve this, a new subset named "t" will be used to store the merged plot. The merging of these two plots is facilitated by the **"arrange_ggplot"** function from the **"metan"** package.

ncol=2: suggested that our two plots will be saved in two different column side by side

After executing the code, you can view the results in the plot area to see the combined plots side by side.

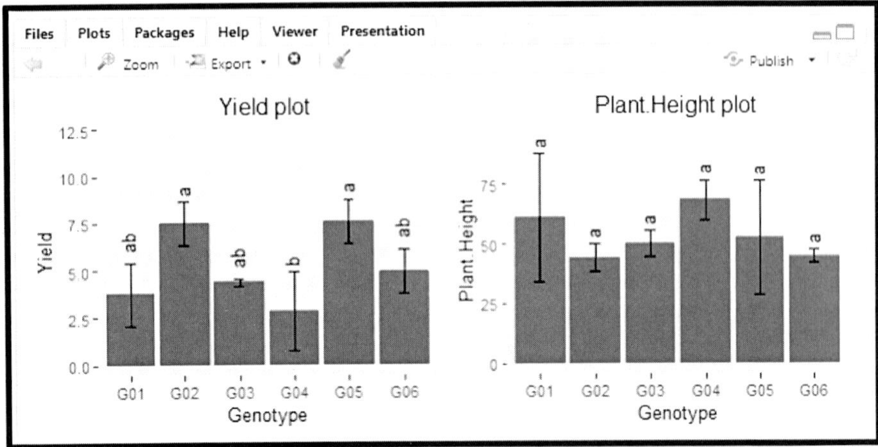


Note: care should be taken at change the name of labels in the ggplot development code for each variable

```
# step 7: Developing bar graph
s=ggplot(data = arrangedtreatments, aes(Genotype, Plant.Height)) +
  scale_y_continuous(limits=c(0,95)) +
  geom_bar(stat= "identity",fill= "Red") +
  geom_errorbar(aes(ymin=Plant.Height-std, ymax=Plant.Height+std), width=0.2) +
  geom_text(aes(label = labels1, y = Plant.Height + std), angle=90, hjust=-0.5, vjust = 0.3) +
  ggtitle("Plant.Height plot") +
  theme(plot.title = element_text(hjust = 0.5)) +
  theme(axis.text.x = element_text(angle = 0))
s
```